WOMEN, LIVESTOCK OWNERSHIP AND MARKETS

This book provides empirical evidence from Kenya, Tanzania and Mozambique, and from different production systems, of the importance of livestock as an asset to women and their participation in livestock and livestock product markets. It explores the issues of intra-household income management and economic benefits of livestock markets to women, focusing on how types of markets, the types of products and women's participation in markets influence their access to livestock income.

The book further analyses the role of livestock ownership, especially women's ownership of livestock, in influencing household food security though increasing household dietary diversity and food adequacy. Additional issues addressed include access to resources, information and financial services to enable women more effectively to participate in livestock production and marketing, and some of the factors that influence this access.

Practical strategies for increasing women's market participation and access to information and services are discussed. The book ends with recommendations on how to mainstream gender in livestock research and development if livestock are to serve as a pathway out of poverty for the poor and especially for women.

Jemimah Njuki is Program Director for a Women and Agriculture Program for CARE USA. She previously led the Gender, Poverty and Impact Group at the International Livestock Research Institute, as well as leading gender mainstreaming and integration of gender in ILRI's research programs.

Pascal C. Sanginga is Senior Program Specialist for Agriculture and Environment at the International Development Research Centre (IDRC) Regional Office for Sub-Saharan Africa, Nairobi, Kenya.

WOMEN, LIVESTOCK OWNERSHIP AND MARKETS

Bridging the gender gap in Eastern and Southern Africa

Edited by Jemimah Njuki and Pascal C. Sanginga

Routledge
Taylor & Francis Group

LONDON AND NEW YORK

earthscan
from Routledge

IDRC | CRDI

International Development Research Centre
Ottawa • Cairo • Montevideo
• Nairobi • New Delhi

ILRI
INTERNATIONAL
LIVESTOCK RESEARCH
INSTITUTE

First published 2013 by Routledge

2 Park Square, Milton Park, Abingdon, Oxfordshire OX14 4RN

711 Third Avenue, New York, NY 10017

Routledge is an imprint of the Taylor & Francis Group, an informa business

First issued in paperback 2018

Co-published with the
International Development Research Centre
PO Box 8500, Ottawa, ON K1G 3H9 Canada
info@idrc.ca / www.idrc.ca
(IDRC published an ebook edition of this book, ISBN 978-1-55250-554-0)

© 2013 **International Livestock Research Institute and the
International Development Research Centre**

British Library Cataloguing-in-Publication Data
A catalogue record for this book is available from the British Library

Library of Congress Cataloging-in-Publication Data
Women, livestock ownership, and markets : bridging the gender gap in
Eastern and Southern Africa / edited by Jemimah Njuki and Pascal C.
Sanginga.
     pages cm
   Includes bibliographical references and index.
   1. Women in development—Africa, Eastern.   2. Women in development—
Africa, Southern.   3. Animal industry—Africa, Eastern.   4. Animal
industry—Africa, Southern.   I. Njuki, Jemimah.   II. Sanginga, P. C.
HQ1240.5.A354W644 2013
305.409676—dc23

                                                          2013011492

ISBN: 978-0-415-63928-6 (hbk)
ISBN: 978-1-138-37710-3 (pbk)

Typeset in Bembo
by Keystroke, Station Road, Codsall, Wolverhampton

CONTENTS

FIGURES

TABLES

CONTRIBUTORS

Juliet Kariuki is a doctoral student at the University of Hohenheim's Institute of Economics and Social Sciences in the Tropics and Subtropics. Past work has focused on understanding marketing and intra-household dynamics of poor men and women in livestock-dependent households in Africa. She also has experience in conducting social research on pastoralists' vulnerability to climate change in arid and semi-arid lands. Her recent publications include work on gender and market analysis, livestock insurance and social impact evaluations.

Samuel Mburu holds a Master of Science degree in Agricultural and Applied Economics, University of Nairobi and a Bachelor of Science degree in Agricultural Economics, Egerton University-Kenya. He has a wide experience in design and execution of research projects, management of massive rural and urban household panel data and data analysis. Recent work includes gender and livestock studies focusing on markets, incomes and implications for food security in Kenya and Tanzania. He is currently involved in the index-based livestock insurance project, working among pastoralists in Kenya and Ethiopia.

Beth Miller is a veterinarian with 20 years of experience integrating livestock development and gender equity in international development. A graduate of Louisiana State University School of Veterinary Medicine in the USA, Dr Miller has owned and operated a small-scale goat dairy, and a mobile private veterinary practice in the USA. For 10 years, she served as Director of Gender Equity for Heifer International, which uses livestock as an entry point for sustainable community development. Dr Miller started Miller Agricultural Consulting Inc. in 2002, to provide project planning, training, research and evaluation services to a variety of clients, including the World Bank, USAID, FAO, ILRI, IFPRI, Heifer International, ADRA, Veterinarians Without Borders, Land O'Lakes and GALVmed.

She is committed to a future where sustainable livestock production contributes to the livelihoods of small-scale farmers and herders, supported by sound agricultural policies.

Bagalwa Nabintu Sanginga currently works as a consultant at the International Livestock Research Institute and holds a Master of Education in Community Development, and a Bachelor of Arts in Rural Development and Regional Planning. She has worked as a consultant for the International Centre for Tropical Agriculture and the Africa Highlands Initiative on gender and livelihood analysis; she also has experience of qualitative data analysis, participatory rural appraisal and baseline studies, and monitoring and evaluation systems of agricultural research for development projects in Eastern and Southern Africa.

Jemimah Njuki holds a PhD in Rural Development and a BSc in Dairy, Food Science and Technology. She has over 15 years' experience working on gender and agriculture, from both a research and development perspective. Currently she is Program Director for a Women and Agriculture Program for CARE USA. She is working on increasing women's productivity and empowerment in equitable agriculture systems at scale in Africa and South Asia. Prior to working with CARE USA, Dr Njuki led the Gender, Poverty and Impact Group at the International Livestock Research Institute, where she also led gender mainstreaming and integration of gender in ILRI's research programs. She was instrumental in the development of the institute's gender strategy and action plan. Prior to joining ILRI, Dr Njuki held various positions at the International Centre for Tropical Agriculture. She has been involved in many efforts in Africa to build capacity on gender and agriculture. She has published widely on gender and agriculture and has co-authored a book, *Innovation Africa: Enriching Farmers' Livelihoods*.

Paula Pimentel has 25 years' experience in agricultural sciences and rural development, covering smallholder and large-scale production. She holds an MSc degree in Animal Production from the University of Pretoria (RSA) and an Honours degree in Veterinary Medicine from Eduardo Mondlane University (UEM-Mozambique). She has vast experience in planning, implementing and evaluating research and extension programs as well as agricultural development projects, country-wide. She has lectured on cattle and small ruminants husbandry for more than 10 years at UEM. She has headed various departments within the government of Mozambique and has served as the Technical Director (nationwide) of Training, Documentation and Technology Transfer (DFDTT) at the Agricultural Research Institute of Mozambique (IIAM). She won the international African Women in Agricultural Research and Development (AWARD) 2010 Fellowship. Paula Pimentel is currently the Senior Agricultural Research and Technology Transfer Advisor at the United States Agency for International Development (USAID), Mozambique Mission.

Pascal Sanginga is currently a Senior Programme Specialist for Agriculture and Environment at the International Development Research Centre (IDRC) based at the Regional Office for Sub-Saharan Africa, Nairobi, Kenya. He is responsible for programming, grant-making and supporting applied research in agricultural development, food security and nutrition. Prior to joining IDRC, he was a senior social scientist with the International Centre for Tropical Agriculture and a research fellow with the Program on Participatory Research and Gender Analysis. Dr Sanginga has co-authored other books including *Innovation Africa: Enriching Farmers' Livelihoods*, and *Managing Natural Resources for Development in Africa: A source book*. Dr Sanginga earned a PhD in Rural Sociology from the University of Ibadan, Nigeria.

Elizabeth Waithanji holds a PhD and an MA in Geography, an MSc in Clinical Studies and a Bachelor of Veterinary Medicine (BVM). Her current work focuses on understanding gender and livestock, looking at the roles of men and women in livestock production in Africa and Asia, and the role that livestock can play as pathway out of poverty, especially for rural women and marginalized communities. Her research focuses mainly on gendered analysis of livestock value chains, inter- and intra-household asset disparities in livestock and food crop production, disparities in market and technology access, and disparities in empowerment among women and men in livelihood intervention projects and the role rights play in effecting the gendered impacts of these interventions. Past work includes a study on mobility and gender relations after sedentarization among Somali nomads of northeastern Kenya and ethnography on gender struggles in development interventions among the Somali of Gedo in southern Somalia.

FOREWORD

Livestock remain a lifeline for many of the world's poorest people. Cattle, goats, sheep, pigs, chickens and other farm animals form part of the livelihood portfolios of an estimated 70 per cent of the world's rural poor women and men. For many smallholder farmers, livestock are essentially four-legged bank accounts, allowing hundreds of millions of "unbanked" poor to save and build assets and to insure themselves against shocks such crop failures, accidents and illnesses. Such "assets" and insurance are particularly important to women, who remain the backbone of global smallholder agriculture, and who are one of the best hopes for ensuring future global food security.

Across the world's varied livestock production systems and regions, women are main actors in poultry, small ruminant and micro livestock production, as well as in dairying, including the processing and marketing of milk and milk products. But women and girls are often still excluded from household decision-making processes, especially regarding the disposal of animals and animal products. This lack of female control over livestock assets and income impinges on family welfare as well as economic growth.

This book provides the first comprehensive analysis of women's ownership of livestock assets, participation in livestock and livestock product markets, intra-household decision-making and income management, and the contribution livestock make to income and food security. The book explores issues of intra-household income management and economic benefits of livestock markets to women, including how the ownership of livestock by women influences household food security, in such areas as the diversity and nutritional adequacy of household meals. The authors discuss practical strategies for increasing women's participation in livestock markets, such as through collective action and better access to livestock information and services. The book concludes with recommendations on how to mainstream gender in livestock research and development work so as better to ensure that livestock serve as a pathway women can take out of poverty.

This book could make a very valuable contribution to the work of practitioners and researchers of gender issues alike. I commend it to you.

Jimmy Smith
Director General, International Livestock Research Institute
January 2013

ACKNOWLEDGEMENTS

This book is a result of the effort of many people. The authors would like to acknowledge the financial contribution of the Ford Foundation, the Collective Action and Property Rights Program (CAPRi) of the International Food Policy Research Institute (IFPRI) and the International Development Research Centre (IDRC), who provided the funding for the research on which this book is based.

The authors acknowledge the great work done by the research teams in Kenya, Tanzania and Mozambique, and the support of the partner organizations including the Sokoine University of Agriculture in Tanzania, the Agricultural Research Institute of Mozambique (IIAM) and the Ministry of Livestock Development, Kenya. We would also like to say thank you to the International Livestock Research Institute (ILRI) regional office for Southern Africa in Mozambique for providing coordination support for data collection in Mozambique, with special thanks to Siboniso Moyo and Felisberto Maute.

A few people provided early comments and reviews on the chapters contained in this book. Special thanks to Nancy Johnson, Cheryl Doss, Agnes Quisumbing, John McDermott and Ruth Meinzen-Dick who, as this book was being written, provided critical comments and insights on gender issues in agriculture and livestock research and development.

Last but not least, thanks to the many men and women farmers and traders who took the time to talk to the teams of researchers. This book would not have been possible without you.

ABBREVIATIONS

ASF	animal source food
BRAC	Bangladesh Rural Advancement Committee
CAPRi	Collective Action and Property Rights Program
CARE	Cooperative for Assistance and Relief Everywhere
DfID	Department for International Development
EADD	East Africa Dairy Development
FAO	Food and Agriculture Organization
FCS	food consumption score
HDDS	household dietary diversity score
HIV/AIDS	human immunodeficiency virus/acquired immune deficiency syndrome
ICRISAT	International Crops Research Institute for the Semi-Arid Tropics
ICT	information and communication technology
IDRC	International Development Research Centre
IFAD	International Fund for Agricultural Development
IFPRI	International Food Policy Research Institute
IIAM	Agricultural Institute of Mozambique
ILO	International Labour Organization
ILRI	International Livestock Research Institute
LINKS	Livestock Information Network and Knowledge System
LSMS	Living Standards Measurement Surveys
M&E	monitoring and evaluation
MAHFP	months of adequate household food provisioning
NAFIS	National Farmers Information Services
NGO	non-governmental organization
ROSCA	Rotating Savings and Loans Association
S&T	science and technology

SEAGA	Socio-economic and Gender Analysis
SEWA	Self Employed Women's Association
SLF	Sustainable Livelihoods Framework
SPSS	Statistical Package for the Social Sciences
TLU	Tropical Livestock Unit
UN	United Nations
UNDP	United Nations Development Programme
USAID	United States Agency for International Development
USD	United States Dollar
VSLA	Village Savings and Loans Association
WDR	World Development Report
WEAI	Women's Empowerment in Agriculture Index
WEF	Women's Empowerment Framework
WFP	World Food Programme

1

GENDER AND LIVESTOCK: KEY ISSUES, CHALLENGES AND OPPORTUNITIES

Jemimah Njuki and Pascal Sanginga

The multiple roles of livestock

The capacity of livestock systems to provide protein-rich food to billions of smallholder rural food producers and urban consumers, generate income and employment, reduce vulnerabilities in pastoral systems, intensify small-scale mixed crop-livestock systems and sustain livelihood opportunities to millions of livestock keepers (ILRI 2012) makes them an appealing vehicle for pro-poor development. Increased consumption of livestock products, particularly in the fast-growing economies of the developing world, has been an important determinant of rising prices for meat and milk (Delgado *et al.* 1999; Delgado 2003). These price surges provide new incentives and opportunities for using livestock as an instrument to help poor people escape poverty due to the multiple benefits that they offer and the multiple roles that they play in different production systems (Rangnekar 1998; Aklilu *et al.* 2008).

Livestock provide income, create employment opportunities and provide food and nutrition security across different production systems and along different value chains. As poor livestock-keeping households tend to be net sellers of livestock products, they benefit from rising livestock prices. Moreover, vulnerable groups, particularly women and the landless, frequently engage in livestock production, thus highlighting the multifaceted virtues of livestock promotion as a pathway out of poverty (Heffernan and Misturelli 2000). Livestock provide a safety net, helping keep poor households from falling into poverty. They are often the only asset women can own/control and can be sold to meet emergency and family health needs.

Livestock also play important roles in securing household food security. This happens through various pathways: (i) in times of food shortages, households sell livestock to purchase other food such as cereals and legumes; (ii) income from regular livestock and livestock product sales is used for food purchases to supplement household food production and to diversify diets; (iii) livestock and livestock products are consumed and provide a protein diet for households.

Importance of livestock to women

Livestock are one of the largest non-land assets in rural asset portfolios, are widely owned by rural households and perform multiple functions. Livestock constitute a popular productive asset with high expected returns through offspring, sale or consumption of products and their use in farming systems. Livestock can also be accumulated (bought) in good times and depleted (sold) in bad times for the purpose of consumption smoothing. Livestock value chains are, however, often more complex than crop value chains, making it difficult to recognize immediate potential entry points for interventions.

In spite of the importance of livestock, a recent review of evidence on the importance of livestock for women by Kristjanson *et al.* (2010) argued that despite two-thirds of the world's more than 600 million poor livestock keepers being rural women (Thornton *et al.* 2003), little research has been conducted in recent years on rural women's roles in livestock keeping and the opportunities livestock-related interventions could offer them. This is in contrast to considerable research on the roles of women in small-scale crop farming, where their importance is widely recognized and lessons are emerging about how best to reach and support them through interventions and policies (e.g. Quisumbing and Pandolfelli 2010; FAO 2011; World Bank 2012).

Livestock have been described as an asset that women can own more easily and that have the potential to contribute to a reduction in the gender asset gap within households (Kristjanson *et al.* 2010). It is often easier for many women in developing countries to acquire livestock assets, whether through inheritance, markets or collective action processes, than it is for them to purchase land or other physical assets or to control other financial assets (Rubin *et al.* 2010). The relative informality of livestock property rights can, however, be disadvantageous to women when their ownership of animals is challenged. Interventions that increase women's access and rights to livestock, and then safeguard the women from dispossession and their stock from theft or untimely death, could help women move along a pathway out of poverty.

Evidence of ownership of livestock by women is, however, scant due to the fact that the collection of sex-disaggregated data has not been common in agricultural surveys. In a review of 72 Living Standards Measurement Surveys (LSMS) and quasi-LSMS by Doss *et al.* (2007), in only three was data collected on individual ownership of farm animals, while the rest assumed that all the livestock was the property of the household, or the head of the household, rather than of the individuals within the household. As a result, most of the comparisons of livestock ownership have been between male- and female-headed households. The few surveys that have looked at individual ownership of livestock have focused on the percentages of households where women have reported owning different species of livestock (Noble 1992; Valdivia 2001). These studies have highlighted the role that small ruminants especially play in securing food, milk protein and cash, and in increasing women's bargaining power. They caution, however, that even in cases where women may have ownership of these species, the marketing and decision-making on the use of money from these assets may still be in the hands of men, thereby undermining the benefits that would be expected to result from women's "ownership". Other qualitative studies provide evidence of the differences in ownership of species, with women more likely to own small stock such as goats, sheep and poultry, and men more likely to own large stock such as cattle and buffalo (Bravo-Baumann 2000; Grace 2007; Heffernan *et al.* 2003; Yisehak 2008).

One of the shortcomings of existing sex-disaggregated livestock data is that it often does not describe information on the value of the livestock but mainly the incidence of ownership of different species and, in a few cases, the actual numbers of different species owned by men and women. Due to the relative value across species, breeds and even age of livestock, understanding the gender disparities or inequalities on livestock ownership, based on this data, is often impossible. As Doss *et al.* (2007) argue, in order to get a better understanding of gender inequalities in asset ownership, it is important to look at both whether women own or don't own livestock, as well as the numbers and value of what they own.

The ownership of livestock and other assets has a bearing on how and who makes decisions on these assets. While some data exists on the relationship between land ownership and agricultural decision-making, this is not the case for livestock. Often, however, these two aspects are not interlinked or followed up in livestock research, which makes it difficult to understand the relationship between ownership and decision-making. Owing to the complexity of ownership, information on rights that individuals, especially women have, over assets is important. For example, data from the Nicaragua LSMS reviewed by Doss *et al.* (2007) showed that although women were sole land owners in 16.3 per cent of households, they made agriculture decisions in only 8.5 per cent of households. Another shortcoming of the current sex-disaggregated data on livestock is the lack of information on the means of acquisition by men and women and how these differ.

This book argues that different livestock and livestock products have different importance for women. It is widely recognized that small livestock such as goats, sheep and poultry are especially important for women. They have more easy access to them, can own them and have control of the animals and their products. While women may not be able to own cattle, in some countries they have control of livestock products (Waters-Bayer 1985; Dieye *et al.* 2005). Women may also benefit more from certain livestock value chains such as local poultry production and marketing, or particular points of value chains such as informal trading, processing or as service providers. In many cases, however, such value chains or segments of value chains where women are found are often low value. Identifying these value chains and increasing their value is critical to increasing women's benefits from livestock production and marketing. An analysis that identifies these points on the value chains, and leads to the selection of interventions that have been used and can be used to increase their value and benefits to women, is crucial. This requires data on the current role of livestock in women's livelihoods, and the challenges and opportunities that women face with regard to acquiring, managing and maintaining livestock.

As a general rule, the degree of commercialization in livestock products is higher than in crops. In all developing countries, livestock add value to resources that have no alternative use, or to on-farm produce. More than in food production, livestock's most important role in food security is to be seen in income generation, starting from the producer down the chain to marketing and processing. Despite this, many interventions on food security seem to focus on crops, with the goal of increasing crop production to ensure food availability. Few interventions or studies have analysed the critical issues of access to food by poor households, or mechanisms by which poor households in predominantly livestock-based systems can increase access to food. Existing evidence suggests that food production and food availability is only part of the problem: as important is food access through increasing income opportunities for the poor. Empirical evidence and studies have provided evidence that poor households spend a significant proportion of their income on food and that livestock is a crucial source of that income. For example in areas of extended poverty and food insecurity, such as the central highlands of Ethiopia, the sale of dung cakes is the most important source of cash income for meeting household food security needs (FAO 1998). Some studies, however, have revealed the tensions and trade-offs between income and food security, as income is likely to be spent on other household needs (education, health, assets and luxuries) and less on food. Gender and intra-household dynamics may influence whether income from sale of livestock is used to meet food security needs or is used for other purposes, thus compromising household food security.

Key gender and livestock issues

Women are major contributors in the agricultural economy, but face various constraints that limit them from achieving optimal livestock production and agricultural development. These constraints include: limited access to land, water and credit; limited information on prices of marketing systems provided by extension agents, which would mean that they find it more difficult to access and maintain profitable market niches and generate more income; limited decision-making powers because of unequal power relations within the household (IFAD 2009). And although women are involved in and may control production, they often do not own the means of production – namely, livestock, land and water (Galab and Rao 2003; Shicai and Jie 2009). Often, too, women lack access to the service and input delivery systems in livestock production, which are male dominated (Sinn et al. 1999; Shicai and Jie 2009). This lack of access and control could be attributed to cultural norms which deny women rights beyond usufruct rights to resources – land, animals and water – and rights to decision-making. A report by the Food and Agriculture Organization (FAO 2011) argues that if women were to have access to the same level of resources as men, agricultural productivity would go up by 10–30 per cent and agricultural output would increase by up to 4 per cent.

Women are more likely to be considered the owners of small livestock compared to larger livestock, and to have a say in the disposal and sale of these and their products, and in the use of income accrued from the sales. Despite their role in livestock production, women's control has traditionally declined when productivity has increased and products are marketed through organized groups such as cooperatives, whose membership is predominantly men (Kergna *et al.* 2010). Studies in the crop sector have shown that the types of products and distance to markets can influence the level of control that women have over these products and the income derived from their sale (Njuki *et al.* 2011).

Compared to crops, little research has been conducted on women's role in livestock farming (Kristjanson *et al.* 2010). The few existing analyses of the role and economic contribution of women to livestock development and the key challenges they face are inconclusive (Niamir-Fuller 1994; Rangnekar 1998; Aklilu *et al.* 2008). This inconclusiveness could be explained, in part, by the fact that the considerable involvement of women in livestock production is underestimated (Sinn *et al.* 1999). For example, most agricultural work is done by women most of whom work for 12–16 hours a day. Moreover, not all women who manage farm resources have access to the income generated by the farm (Sinn *et al.* 1999). In addition, of all rural agricultural extension services, women are able to access only 5 per cent of what men access (FAO 1996–2001). Other likely explanations as to why research regarding the role and economic contribution of women to livestock development and the key challenges they face is inconclusive include the fact that gender roles, relations and ideologies are not studied prior to and during interventions involving women and livestock; and attitudes and values regarding livestock, between men and women, are highly polarized (Kristjanson *et al.* 2010).

The International Livestock Research Institute (ILRI) recently developed a conceptual framework on livestock as a pathway out of poverty. This framework takes a "livelihoods approach" that centralizes the importance of assets, markets and other institutions. In a literature review, Kristjanson *et al.* (2010) apply the framework using a gender lens to discuss livestock as pathway out of poverty for women. The authors hypothesized that livestock pathways out of poverty are: (i) securing current and future assets; (ii) sustaining and improving the productivity of agricultural systems in which livestock are important; and (iii) facilitating greater participation of the poor in livestock-related markets. These three pathways are well emphasized in this book.

About this book

This book provides empirical evidence from sex-disaggregated data collected in Kenya, Tanzania and Mozambique, and from different production systems, of the importance of livestock as an asset to women and their participation in livestock and livestock product markets. It explores the issues of intra-household income management and economic benefits of livestock markets to women, focusing on how types of markets, the types of products and women's participation in markets influence their access to livestock income. The book further analyses the role of livestock ownership, especially women's ownership of livestock, in influencing household food security through increasing household dietary diversity and food adequacy. Additional issues addressed include access to resources, information and financial services to enable women more effectively to participate in livestock production and marketing, and some of the factors that influence this access. Practical strategies for increasing women's market participation and access to information and services are discussed. The book ends with recommendations on how to mainstream gender in livestock research and development if livestock are to serve as a pathway out of poverty for the poor, and especially for women.

The book focuses on a few critical questions:

- What are the patterns of livestock ownership and what is the importance of livestock as an asset for women?
- What livestock, livestock products and markets have the greatest benefits for women? What are the patterns of market participation? And are these dependent on the livestock species or products?
- How do these patterns of market participation influence income management by women? Does the type of livestock, product and markets they are sold to influence whether income will be managed by men, women or jointly?
- What are the different pathways through which livestock improve household food security? What roles do livestock and livestock products managed by women play in coping with food vulnerabilities?

Each of the chapters of the book is dedicated to these questions. Chapter 2 focuses on the methodology used in the collection and analysis of the data, describing the quantitative and qualitative methods used and the analysis employed. The chapter starts with a description of the household model that provided the theoretical rationale for the data collection procedures used. Chapter 3 looks at patterns of livestock ownership, the contribution of livestock to women's, men's and household asset portfolios, and narrows down to focus on women's ownership of livestock and decision-making. Chapter 4 uses a combination of qualitative and quantitative data to understand patterns of market participation across species and products by men and women, as well as the market preferences they have and reasons for these preferences. Chapter 5 analyses intra-household income management and the factors that influence whether income from livestock will be managed by men or by women, or jointly. The analysis looks at the species differences and product differences to identify patterns and groups of products or species where women manage more income compared to others. For poor rural men and women to access markets and accumulate assets, access to information and financial services, including savings, is crucial. Chapter 6 focuses on gender differences in access to livestock production and marketing information, and the common sources of this information for men and women. It goes on to look at access to credit, different uses of credit by men and women, as well as access to savings. The last analysis chapter (chapter 7) focuses on the role of livestock in improving food security, and especially looking at women's ownership of livestock and how that influences dietary diversity and the consumption of animal source foods. Chapter 8 discusses some practical strategies for how to mainstream gender in livestock research and development if livestock are to serve as a pathway out of poverty for the poor, and especially for women. The book concludes with a summary chapter on some key issues, findings, conclusions and the implications of these results for livestock development research, policies and programs.

References

Aklilu, H. A., Almekinders, C. J., Udo, H. M. and Van der Zijpp, A. J. (2008) Village poultry consumption and marketing in relation to gender, religious festivals and market access. *Tropical Animal Health Production* 39: 165–177.

Bravo-Baumann, H. (2000) *Gender and Livestock: Capitalisation of Experiences on Livestock Projects and Gender.* Working document, Swiss Agency for Development and Cooperation, Bern.

Delgado, C. (2003) Rising consumption of meat and milk in developing countries has created a new food revolution. *Journal of Nutrition* 133(11): 3907S–3910S.

Delgado, C., Rosegrant, M., Steinfeld, H., Ehui, S. and Courbois, C. (1999) *Livestock to 2020: The Next Food Evolution.* Food, Agriculture and the Environment Discussion Paper 28. Washington, DC: IFPRI.

Dieye, P. N., Ly, C. and Sane, F. C. N. (2005) *Etude des service d'élevage dans la filière laitere au Sénégal.* Pro-Poor Livestock Policy Initiative (PPLPI) Internal Report. Rome: FAO.

Doss, C., Grown, C. and Deere, C. D. (2007) *Gender and Asset Ownership: A Guide to Collecting Individual-level Data.* World Bank Policy Research Working Paper 4704. Washington, DC: World Bank.

FAO (1996–2001) *Plan for Action for Women in Development.* Rome: FAO.

FAO (1998) *Crop and Food Supply Assessment Mission to Ethiopia.* Rome: FAO Global Information and Early Warning System on Food and Agriculture, World Food Programme.

FAO (2011) *The State of Food Agriculture: Women and Agriculture, Closing the Gender Gap for Development.* Rome: FAO.

Galab, S. and Rao, C. (2003) Women self-help groups: poverty alleviation and empowerment. *Economic and Poverty Weekly* 38(12/13): 1274–1283.

Grace, D. (2007) *Women's Reliance on Livestock in Developing-country Cities.* ILRI Working Paper. Nairobi: ILRI.

Heffernan, C. and Misturelli, F. (2000) *The Delivery of Veterinary Services to the Poor: Preliminary Findings from Kenya.* Report for DfID. Reading: Veterinary Epidemiology and Economics Research Unit, University of Reading.

Heffernan, C., Misturelli, F. and Pilling, D. (2003) *Livestock and the Poor: Findings from Kenya, India and Bolivia.* London: Animal Health Programme, Department for International Development.

IFAD (2009) *Gender and Livestock: Tools for Design.* Rome: IFAD. Available at: http://www.ifad.org/lrkm/events/cops/papers/gender.pdf (accessed 6 February 2011).

ILRI (2012) *Livestock Matter(s): Where Livestock Can Make a Difference*. ILRI Corporate Report 2010–2011. Nairobi: ILRI.

Kergna, A., Diarra, L., Kouriba, A., Kodoi, B., Teme, B. and McPeak, J. (2010) Role of farmer organizations in the strategy for improving the quality of life for livestock producers in Mali. Research Brief 10-02-MLPI. Global Livestock CRSP (Cooperative Research Program), January.

Kristjanson, P., Waters-Bayer, A., Johnson, N., Tipilda, A., Njuki, J., Baltenweck, I. *et al.* (2010) *Livestock and Women's Livelihoods: A Review of the Recent Evidence*. ILRI Discussion Paper No. 20. Nairobi: ILRI.

Niamir-Fuller, M. (1994) *Women Livestock Managers in the Third World: Focus on Technical Issues Related to Gender Roles in Livestock Production*. Staff Working Paper 18, Rome: IFAD.

Njuki, J., Kaaria, S., Chamunorwa, A. and Chiuri, W. (2011) Linking smallholder farmers to markets, gender and intra-household dynamics: does the choice of commodity matter? *European Journal of Development Research* 23: 426–433.

Noble, A. (1992) Women, men, goats, and bureaucrats: the Samia women's dairy goat project. In C. M. McCorkle (ed.) *Plants, Animals, and People*. Boulder, CO: Westview Press.

Quisumbing, A. and Pandolfelli, L. (2010) Promising approaches to address the needs of poor female farmers. *IFPRI Note* 13: 1–8.

Rangnekar, S. (1998) Women in livestock production in developing countries. Paper presented at the International Conference on Sustainable Animal Production, 24–27 November, Hisar, India.

Rubin, D., Tezera, S. and Caldwell, L. (2010) *A Calf, a House, a Business of One's Own: Microcredit, Asset Accumulation, and Economic Empowerment in GL CRSP Projects in Ethiopia and Ghana*. Washington, DC: Global Livestock Collaborative Research Support Program.

Shicai, S. and Jie, Q. (2009) Livestock projects in southwest China: women participate, everybody benefits. *Leisa Magazine* 25(3 Sept).

Sinn, R., Ketzis, J. and Chen, T. (1999) The role of women in the sheep and goat sector. *Small Ruminant Research* 34(3): 259–269.

Thornton, P. K., Kruska, R. L., Henninger, N., Kristjanson, P. M., Reid, R. S. and Robinson, T. P. (2003). Locating poor livestock keepers at the global level for research and development targeting. *Land Use Policy* 20(4): 311–322.

Valdivia, C. (2001) Gender, livestock assets, resource management, and food security: lessons from the SR-CRSP. *Agriculture and Human Values* 18(1): 27–39.

Waters-Bayer, A. (1985) *Dairying by Settled Fulani Women in Central Nigeria and Some Implications for Dairy Development*. ODI Pastoral Development Network Paper 20c. London: Overseas Development Institute.

World Bank (2012) *World Development Report 2012: Gender Equality and Development*. Washington, DC: World Bank.

Yisehak, K. (2008) Gender responsibility in smallholder mixed crop–livestock production systems of Jimma zone, South West Ethiopia. *Livestock Research for Rural Development* 20(11).

2

COLLECTING AND ANALYSING DATA ON INTRA-HOUSEHOLD LIVESTOCK OWNERSHIP, MANAGEMENT AND MARKETING

Jemimah Njuki, Elizabeth Waithanji and Samuel Mburu

Collecting data on gender and intra-household dynamics: household models

For a long time, households were assumed to behave as one decision-making unit (the unitary model of the household). The unitary model views the household as a single *economic unit* that works as a group for its own good and all members of the household contribute in an *altruistic* manner towards the benefit and functioning of the entire household (Katz, 1996; Fortin and Lacroix 1997). This approach has, however, been found to have methodological, empirical and welfare economic limitations (Vermeulen 2002). A valuable alternative to this traditional unitary model is the collective approach to household behaviour which takes account of the fact that households consist of different members who go through an intra-household bargaining process in the allocation of resources and decision-making.

A synthesis (Quisumbing 2003) of literature on household decision-making summarizes overwhelming evidence from empirical case studies from several countries in different contexts that households do not act as unitary model when making decisions. The synthesis supports a non-unitary model of household decision-making.

There are two types of collective household models: cooperative and non-cooperative. These are illustrated in Figure 2.1.

In the non-cooperative model, each household member acts in order to maximize his or her own utility while in the cooperative model the households act as a unit to maximize the welfare of members. The analysis presented in this book uses an adapted collective cooperative model as its theoretical basis and the data collection and the analysis is grounded in this. This adapted model assumes collective behaviour in which household members, in this case the male and female adults

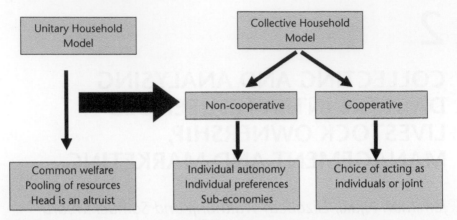

FIGURE 2.1   An illustration of the unitary and collective bargaining models

within the household, may choose to act individually or jointly. This has implications both for the data collection and analysis.

Collecting data from individuals within the household

The gender dimensions of livestock ownership, participation in markets, and access to information and technologies has often been collected at household level (Doss *et al.* 2007). There are, however, some exceptions especially in Asia (Quisumbing and de la Brière 2000; Kumar and Quisumbing 2010). As a result, most of the comparisons of livestock ownership, participation in markets and other important variables have been between male- and female-headed households. This type of analysis has masked intra-household dynamics and decision-making processes that have important implications for women, and for households. Data collection is also often targeted at heads of households, with the assumption that he (typically) is the owner of livestock and other assets. Rarely is data collection on asset and livestock ownership targeted at other individuals within households, and especially women.

Collecting intra-household resource allocation, income and decision-making data is complex. Studies have collected individual data (from both men and women within households) mainly on assets and access to resources. Doss *et al.* (2007), in reference to individual data on assets and resources, point out that such individual data is important to understand the relationship between men and women, the types of assets and, in this case, sources of income that they manage – and it can provide a measure of intra-household inequality that can be used across countries and regions. Collection of this data, however, should go beyond data on men and women, but should be collected from both men and women. This adds another complexity to data collection. In many countries and regions, due to cultural complexities, men may not be willing to allow their spouses to be interviewed, especially by people who are not from the local community or by male data collectors. This means that there has to be a negotiation process to allow women to be interviewed and to be interviewed away from men to reduce men's influence on data and information given by women. Another complexity arises due to inconsistencies in data between men and women, especially when the same questions are asked of men and women, for example questions on how much money is made from the sale of a particular commodity and how much is managed by the man and how much is managed by the woman. Depending on who is asked, the information may be different. This leads to more follow-up discussions and clarifications than there would otherwise be in an ordinary interview where one member is interviewed on behalf of the household. These follow-ups and discussions can, however, lead to better quality of data.

The study used different strategies to interview both men and women and to get around these complexities, including: (i) the use of both male and female enumerators; (ii) starting women's interviews with questions on domains that they have control over within the household such as food security and nutrition before moving on to more sensitive and complex questions of income and income management; (iii) multiple visits to households for follow-up discussions; (iv) thorough comparison of men and women's responses to similar questions.

"Individual" vs "jointness": understanding context

The authors recognize that gender norms are complex and dynamic. They change gradually, in response to shifting economic, political and cultural forces that can create new constraints and opportunities for women (Quisumbing and Pandolfelli 2008). Meanings also differ, depending on the cultural context. Kabeer's (2001) analysis of credit interventions has attributed the different judgements of successes and failures of the interventions to two main explanations; first, there exist different understandings of intra-household power relations. For example, "joint management" could be a disguised male dominance. Second, there is no clear definition of what empowerment entails, and often the context in which the definition is made is obscure. For example, empowerment can be manifested differently in contexts of conflict and cooperation, and both empowerment and disempowerment may manifest as autonomy, dependence or interdependence within the household. There needs to be a nuanced understanding, during the data collection and analysis, of these meanings and their implications.

Framing gender research questions guiding data collection

Table 2.1 summarizes the key research questions that this book addresses and the type of data collected to answer these questions. It also gives a summary of some of the tools used to collect the different types of data.

Methods for data collection

Quantitative data was collected through household surveys. The questionnaire had two separate modules, one administered to male adults and the second to the female

TABLE 2.1 Key research questions and type of data collected

Key research question	Type of data collected	Tools used
What are the patterns of livestock ownership and what is the importance of livestock as an asset for women?	Ownership of different livestock species by different members of the household, ownership of other assets	Household survey
What livestock, livestock products and markets have the greatest benefits for women? What are the patterns of market participation? Are these dependent on the livestock species or products?	Men's and women's market preferences, markets where livestock and livestock products are sold and who in the household sells	Scoring techniques Household survey Market chain maps
How do these patterns of market participation influence income management by women? Does the type of livestock, product and markets they are sold to influence whether income will be managed by men, by women or jointly?	Market participation, income control by women, household dietary diversity, household food adequacy, coping strategies during food shortage	Household survey
What is the role of livestock in improving food security and as a coping mechanism during food deficit periods for vulnerable households?	Household dietary diversity, household food adequacy, consumption of animal source foods, coping strategies during food shortage	Household survey

adults in male-headed households. The main module contained all the household-level information, including household membership as well as asset, income and decision-making modules. The asset, income and decision-making modules were then asked to women in the second module.

As much as possible, women were interviewed away from men so as to increase the likelihood of capturing their objective perspectives of asset ownership and decision-making. Two strategies were employed to achieve this: (i) using two enumerators, a male and a female for each household and (ii) starting the female adult questionnaire with the food security questions. Often a combination of these strategies worked to get independent data from men and women. In the first strategy, the male adult and female adult were interviewed simultaneously and separately. When this did not work, starting off the female adult interview with the food security questions convinced men that the women would not contradict them as it seemed their questions were related to a domain that men had less knowledge of, that is, food consumption patterns of household members, especially children.

Asking men and women from the same household the same question adds a meaningful and unique dimension to this study. Empirical evidence shows that men and women from the same households do not operate as a single unit, but instead individual household members are likely to have different objectives and thus are likely to function independently. At the same time, individual household members can also choose to function jointly. Therefore, intra-household studies can reflect various household dynamics, including those on decision-making, income control and expenditure patterns. Data was analysed at the aggregate level and compared across men and women for the variables described. Qualitative data was collected mainly though focus group discussions and key informant interviews.

Sampling was multi-staged with a purposive selection of districts based on three criteria: presence of multiple livestock species; availability of livestock marketing groups; and production system. For the household surveys, the sampling frame was the smallest administrative unit from which the livestock keepers to be interviewed resided, often a sub-location or village in densely populated areas. Except for Mozambique, there was no existing list of farmer households in the study sub-locations; therefore a comprehensive list of all households was compiled with local elders, administrators and the group officials.

Data was collected from a total of 730 households. A total of 1332 interviews were conducted. In male-headed households, both the male adult and female adult were interviewed. Male-headed households constituted 81.5 per cent and female-headed households the remaining 19.5 per cent of the sample. The proportion of female-headed households ranged from 15.2 per cent in Mozambique to 25.5 per cent in Kenya.

Locations and site characteristics

The research reported in this book was conducted in Tanzania, Kenya and Mozambique. In all countries, districts were stratified by agricultural potential, production system and/or market access (see Table 2.2). In Tanzania, data was collected from five districts – Kilombero, Kibaha, Gairo, Mvomero and Morogoro. The sites visited are all from mixed crop-livestock systems and, except for Kilombero, all sites are categorized as having high agricultural potential. Kibaha, Gairo and Mvomero are classified as having good market access, with not more than four hours of travel to the nearest market, while Kilombero and Morogoro Rural have low market access, with more than four hours travel to market.

In Kenya, the data was collected in four districts – Kajiado and Tharaka, which are classified as semi-arid, and Meru and Kiambu, which have high agricultural potential. Apart from Tharaka, where the production system is livestock-based, all the sites are classified as having mixed crop-livestock systems. Kajiado and Kiambu have good market access while Meru and Tharaka have low market access.

In Mozambique, data was collected in nine villages from Gaza Province, across two administrative posts (Chicualacuala Sede and Mapai), and one location (Chidulo) in one district, Chicualacuala. All the sites are considered semi-arid, with livestock as the main production system and poor market access.

TABLE 2.2   Description of sites in Tanzania, Kenya and Mozambique

Tanzania

District	Agricultural potential	Production system	Market access
Kibaha	High potential	Mixed crop–livestock	Mostly good (less than four hours to market). Dar es Salaam market has unlimited demand for commercial chickens
Gairo	High potential	Mixed crop–livestock	Mostly good (less than four hours to market). Morogoro market has more limited demand than Dar es Salaam
Kilombero	Low potential	Mixed crop–livestock	Mostly poor (more than four hours to market). Demand from local markets is limited
Morogoro Rural	High potential	Mixed crop–livestock	Mostly poor (more than four hours to market)
Mvomero	High potential	Mixed crop–livestock	Mostly good (less than four hours to market)

Kenya

District	Agricultural potential	Production system	Market access
Kiambu	High potential	Mixed crop–livestock	Mostly good (less than four hours to market). Nairobi market has unlimited demand
Meru	High potential	Mixed crop–livestock	Mostly good (less than four hours to market). Meru market has more limited demand than Nairobi
Kajiado	Semi-arid	Mixed crop–livestock	Mostly poor (more than four hours to market). Ngong area farmers have easy access to the Nairobi market
Tharaka	Semi-arid	Mainly livestock	Mostly poor (more than four hours to market)

Mozambique

Administrative post/location	Agricultural potential	Production system	Market access
Chicualacuala Sede	Semi-arid	Mainly livestock	Mostly poor
Mapai	Semi-arid	Mainly livestock	Mostly poor
Chidulo	Semi-arid	Mainly livestock	Mostly poor

Various commodity livestock value chains were reviewed across the three countries. In Tanzania, data was collected on three different commodity value chains: dairy goats (indigenous and Norwegian cross) for breeding and milk; indigenous and exotic chickens for meat and eggs; and bees for honey and wax. In Kenya, data was collected on four different commodity value chains: dairy cattle (mainly crossbred cattle); dairy goats (pure Toggenburg and their crosses); indigenous and exotic chickens for meat and eggs; and bees for honey and wax. In Mozambique, data was collected on three different commodity value chains, namely: cattle, goats and chickens. Marketing of livestock products such as milk and eggs was very low in Mozambique.

Description of households

There were several distinct differences across the three countries (see Table 2.3). The proportion of female-headed households was highest in Kenya and lowest in Mozambique. Similarly, the highest proportion of heads of households with primary education was in Kenya and lowest in Mozambique. On the other hand, Mozambique had the highest proportion of households keeping cattle and goats while Tanzania had the lowest proportion keeping cattle and Kenya had the lowest proportion keeping goats. Kenya had the highest proportion of households that had at least one member belonging to a group (88.4 per cent), while Mozambique had the lowest at 28.7 per cent. Average size of land holding in Mozambique was significantly higher than in Kenya and Tanzania. Average ages of the heads of households did not differ significantly across the three countries.

Data analysis

The analysis used several exploratory tools and methods. For quantitative data, descriptive statistics, including proportions, comparisons of means and chi square tests were used especially to make comparisons between men and women, between species and between countries. Some of the calculations made to compare livestock

TABLE 2.3 Characteristics of the sampled households by country

	Kenya	*Tanzania*	*Mozambique*
Number of households sampled	243	237	250
% of female-headed households	25.5	17.7	15.2
Average age of head of households	54.1	48.5	48.6
% with primary education and above	83.2	77.2	42.5
% keeping cattle	53.3	21.9	74.7
% keeping goats	32.5	40.5	73.5
% keeping chickens	60.5	77.2	76.3
% belonging to a group	88.4	57.8	28.7
Average land holding (ha)	1.26	2.50	4.56

across species, and to analyse the contribution of livestock to women's and household total asset portfolios are described below.

Calculating Tropical Livestock Units

Owing to the differences in value and ownership patterns of different livestock species, it has been difficult to compare livestock holdings of men and women in real terms. In order to do this, the concept of an "Exchange Ratio" has been developed, whereby different species of different average size can be described by a common unit and compared; this unit is the Tropical Livestock Unit (TLU). While this is a useful comparative measure, different livestock species have different importance for women. For example the ownership of chickens and goats might be preferred by women as they can make decisions on, and control and manage products and incomes from these species.

The calculation of the TLU uses the sub-Saharan Africa values recommended by the FAO (2002). Table 2.4 shows commonly used definitions of TLUs in sub-Saharan Africa. This version of the TLU does not account for breed and feed system differences and has been recommended only for generalized analysis as that done in this book.

$$\text{Total livestock holding} = \sum_{i=1}^{n} TLU_i$$

where n = number of species/type, TLU_i = TLU for species/type i.

Calculating the asset index

One of the key gaps in the evidence of the importance of livestock to women is the overall contribution of livestock to women's asset portfolio. In this analysis, we focus only on movable assets, including domestic and farm assets but excluding land and

TABLE 2.4 Tropical Livestock Units (TLU) conversion rates

Species (animal type)	TLU equivalent
Cattle – oxen/bull	1.0
Cattle – local cow	0.8
Cattle – heifers	0.5
Cattle – immature males	0.6
Cattle – calves	0.2
Sheep/goats	0.1
Horses	0.8
Camel	1.1
Donkeys/mules	0.5
Poultry	0.01

housing. The rationale for excluding the land and housing was due to the complexity of obtaining ownership and claims to ownership of the land owing to the customary nature of tenure systems in the three countries, which also applies to the housing. Both men and women were better able to describe ownership of the movable assets.

TABLE 2.5 Weight and age adjustments for calculating the asset index

Asset (g)	Weight of asset (ω_g)	Age (adjustment for age shown in cell) (a)		
		< 3 yrs old	3–7 yrs old	> 7 yrs old
Animal		Calves	Immature male/heifer	Adult
Cattle	10	× 0.2	× 0.5	× 1
Horses	10			
Sheep/goats	3			
Poultry	1	no adjustment		
Pigs	2			
Domestic assets		< 3 yrs old	3–7 yrs old	> 7 yrs old
Cooker	2			
Kitchen cupboard	2			
Refrigerator	4			
Radio	2			
Television	4			
DVD player	4	× 1	× 0.8	× 0.5
Cell phone	3			
Chairs	1			
Mosquito nets	1			
Gas stove	2			
Transport		< 3 yrs old	3–7 yrs old	> 7 yrs old
Car/truck	160			
Motorcycle	48			
Bicycle	6	× 1	× 0.8	× 0.5
Cart (animal drawn)	12			
Productive		< 3 yrs old	3–7 yrs old	> 7 yrs old
Hoes	1			
Spades/shovels	1			
Ploughs	4			
Treadle pump	6	× 1	× 0.8	× 0.5
Powered pump	12			
Sewing machine	4			

Asset indices for different assets, including livestock, were developed with the asset index methodology developed by the Bill and Melinda Gates Foundation (2010) for the evaluation of their agriculture programs. The asset index was calculated for all movable assets including livestock. Each of the assets was assigned a weight (ω) and then adjusted for age. The weight is calculated based on the value of the asset compared across countries. This ensures that assets of the same value are accorded the same weight, despite country or location differences in prices.

$$\text{Household Domestic Asset Index} = \sum_{g=1}^{G}\left[\sum_{i=1}^{N}(\omega_{gi} \times a)\right],$$

$$i = 1, 2, \ldots, N; g = 1, 2, \ldots, G$$

where, ω_{gi} = weight of the i'th item of asset g, N = number of asset g owned by household, a = age adjustment to weight, G = number of assets owned by household.

Probit analysis

The probit analysis has been used for several of the analyses presented in the next few chapters to test the probability of occurrence and in cases where the *dependent variable* took two values, that is, a *binary response model*. For example it is used in chapter 3 to explore the probability of women owning livestock. In this case the dependent variable took a binary form:

1 = women in the household owned livestock and
0 = women in the household did not own livestock

The probit model took the form:

$$P\gamma(Y = 1 \mid X) = \emptyset(X'\beta)$$

Where:

$P\gamma$ denotes the probability of women owning or not owning livestock (1 or 0)
X is a vector of regressors on the spouse's and household characteristics
$\emptyset$ is the Cumulative Distribution Function (CDF) of the standard normal distribution
β is a parameter typically estimated by maximum likelihood.

Linear regressions

Other analyses use the linear regression to model the relationship between a scalar dependent variable y and one or more explanatory variables denoted by X.

For example, in chapter 3, a linear regression model to analyse factors influencing women's ownership using the TLUs owned by women as the dependent variable is used and takes the form:

$$Y_i = \beta_0 + \beta_i X_i + \varepsilon_i$$

Where Y_i denotes the TLU belonging to women and $X_i * X$ and $_{i*}$ are the independent variables.

Conclusion

Given the different roles that men and women play in agriculture, the different access to and ownership of resources, and the different impact agriculture interventions have on men and women, the collection of sex-disaggregated data should be the norm rather than the exception. Collecting sex-disaggregated data goes beyond the stratification of households as male- and female-headed households and should take into account the intra-household access to and ownership of resources and decision-making. A systematic process for doing this should include framing the gender questions as part of the design of the research, developing tools with sex disaggregation of the key indicators of interest, collecting information from both men and women, and analysing the data to understand gender differences and similarities. While this may imply additional resources and capacity-building for research staff, there is a tremendous pay-off when sex-disaggregated data is used to inform policy and programs on interventions that have potential to reduce gender disparities.

References

Bill and Melinda Gates Foundation (2010) *Agricultural Development Outcome Indicators: Initiative and Sub-initiative Progress Indicators and Pyramid of Outcome Indicators.* Seattle, WA: Bill and Melinda Gates Foundation.

Doss C., Grown, C. and Deere, C. D. (2007) *Gender and Asset Ownership: A Guide to Collecting Individual-level Data.* Policy Research Working Paper 4704. Washington, DC: World Bank.

FAO (2002) *What are Tropical Livestock Units? Livestock and Environment Toolbox.* Rome: FAO.

Fortin, B. and Lacroix, G. (1997) A test of the unitary and collective models of household labour supply. *Economic Journal* 107(443). Available at: http://www.jstor.org/stable/2957843 (accessed May 2013).

Kabeer, N. (2001) Conflicts over credit: re-evaluating the empowerment potential of loans to women in Bangladesh. *World Development* 29(1): 63–84.

Katz, E. (1996) Intra-household economics: neo-classical synthesis or feminist-institutional challenge? Mimeo, Dept. of Economics, Barnard College, USA.

Kumar, N. and Quisumbing, A. (2010) *Does Social Capital Build Women's Assets? The Long-term Impacts of Group-based and Individual Dissemination of Agricultural Technology in Bangladesh.* CAPRi Working Paper No. 97. Washington, DC: IFPRI. Available at: http://dx.doi.org/10.2499/CAPRiWP97 (accessed May 2013).

Quisumbing, A. R. (2003). What have we learned from research on intra-household allocation? In A. Quisumbing (ed.) *Household Decisions, Gender, and Development: A Synthesis of Recent Research.* Washington, DC: IFPRI.

Quisumbing, A. and de la Brière, B. (2000) *Women's Assets and Intrahousehold Allocation in Rural Bangladesh: Testing Measures of Bargaining Power.* FCND Discussion Paper. Washington, DC: IFPRI.

Quisumbing, A. and Pandolfelli, L. (2008) Promising approaches to address the needs of poor female farmers. *IFPRI Note* 13: 1–8.

Vermeulen, F. (2002) Collective household models: principles and main results. *Journal of Economic Surveys* 16(4): 533–564.

3

GENDER AND OWNERSHIP
OF LIVESTOCK ASSETS

Jemimah Njuki and Samuel Mburu

Background

Asset ownership is often highly correlated with economic growth, poverty reduction and with a reduction to vulnerability and risk at the household level (Barham *et al.* 1995; Banerjee and Duflo 2003; Birdsall and Londono 1997; Deere and Doss 2006). There is increasing evidence that women's absolute and relative asset levels are important to development outcomes, directly through their influence on decision-making and indirectly by conditioning women's ability to participate in and benefit from specific livelihood strategies, development programs, etc. Livestock are thought to be one of the most important assets for women as they are a productive asset that they can easily own and that are not bound by complex property rights compared to, for example, land. There is, however, little evidence available on the extent to which women own livestock, which species are most important to them, how they acquire livestock, or how important livestock are relative to other assets, for women and for their households.

This chapter provides a framework for analysing the role of livestock as an asset for women, using an asset index to analyse the gender asset disparity in households and the contribution of livestock to men's, women's and joint household assets in Kenya, Tanzania and Mozambique. The first strand of analysis focuses on the patterns of livestock ownership across species by men and women within households. While these patterns are illuminating in understanding livestock ownership within the household, it does not allow for the comparison of total livestock asset portfolios owned by men and women or jointly and therefore the relative value of their livestock asset portfolio compared to other assets. The second analysis therefore uses Tropical Livestock Units (TLUs) to compare the relative value of livestock owned by men, women and jointly, and an asset index to analyse the contribution of livestock to men's and women's total asset portfolios. The chapter recognizes that ownership of livestock by women, however, does not always imply that they have ultimate control of this livestock. Women may own livestock, acquired through the market or inheritance before or during marriage, but may not have decision-making authority over such livestock. The third level of analysis, therefore, looks at decision-making on women-owned livestock. The fourth strand of analysis looks at how women acquire livestock and the factors that influence the owner-ship of livestock by women. The results presented provide a better understand-ing of the potential role for livestock in improving women's welfare, as well as which types of strategies, inside and outside the livestock sector, are likely to have the biggest impact on empowering women and reducing gender asset disparities.

Some studies (Doss *et al*. 2007; Torkelsson and Tassew 2008) have interviewed or recommend interviewing both men and women within households to collect data on ownership and access to resources. This is useful in order to disentangle the social context of ownership and the differences in perceptions regarding asset ownership and control. To capture the intra-household issues of asset ownership, only data from households with a male adult and a female adult was included in the analysis.

Gender and assets: what do we know?

Assets, in this chapter, are defined as stocks of financial, human, natural and social resources that can be acquired, developed, improved and transferred across genera-tions (Ford Foundation 2004). The Sustainable Livelihoods (SL) Framework identifies five types of capital which could be related to assets. These include social, financial, physical, natural and human capital (Carney *et al*. 1999; DfID 1997; Scoones 2009). Access to and ownership of assets within and beyond the household is critical for increasing agricultural productivity and enabling people to move out of poverty (Doss *et al*. 2011). Assets have been classified in various ways; for example Doss *et al*. (2007), in their guide to collecting individual-level data on assets, focus on physical and financial assets and group them into land, livestock, housing, non-farm business assets, financial assets including savings, pensions and bonds, as well as other physical assets such as domestic furniture and farm equipment. Meinzen-Dick *et al*. (2011) have developed a conceptual framework that looks at the different assets, the gendered nature of these assets and the links to livelihood outcomes and welfare impacts. This framework is shown in Figure 3.1.

While many studies have looked at household ownership of assets as measures of wealth, the gender dimensions of asset ownership and their implications have not been studied as well, due to a lack of awareness of the gender asset distribution as well as of empirical information and data on intra-household ownership, as most assets data is collected at household level (Doss *et al*. 2007). There are, however, some exceptions especially in Asia (see Kumar and Quisumbing 2010; Quisumbing and de la Brière 2000). Within the household or a family, women may not necessarily share in the wealth of men (Deere and Doss 2006). Decisions may be made, and

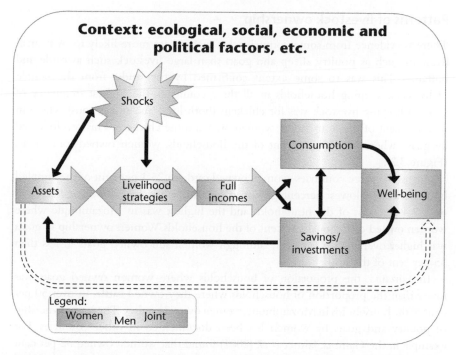

FIGURE 3.1 Schematic representation of a gendered livelihood conceptual framework

sometimes wealth distributed, within a continuum with consensus of members of the family on one end or a dominant single member – the benevolent dictator – on the other end (Lundberg and Pollak 1996; Marchant 1997). Where research on intra-household gender asset distribution has been conducted, women have been shown to own assets of a significantly lower value, with less money in their individual accounts, than men (Antonopoulos and Floro 2005). These asymmetries in gender asset ownership within the household justify the need for gendered asset research even though such research poses a methodological challenge.

Reducing the gender asset gap or putting assets in the hands of women has been shown to have positive outcomes, not only for women themselves but for households. Women's ownership of assets has been shown to increase their bargaining power (Friedemann-Sánchez 2006), their role in household decision-making (Agarwal 1998, 2002; Mason 1998) and expenditures on children's education and health (Allendorf 2007; Duraisamy 1992; Quisumbing 2003; Quisumbing and Maluccio 2000). The gender asset gap is also a critical indicator of women's empowerment and has been recommended for use as an indicator of progress towards achievement of Millennium Development Goal 3 (Grown et al. 2005). It provides a better measure of gender inequality and women's economic empowerment compared to use of such indicators as income. The gender asset disparity is caused by many factors, including social norms, intra-household differences in access, market conditions and government policies.

Patterns of livestock ownership

There is evidence from some countries that women are more likely to own small livestock such as poultry, sheep and goats than large livestock such as cattle and buffaloes. This was to some extent confirmed by the study: from the sample of livestock-keeping households in all three countries, the highest frequency of women keeping livestock was for chickens (both local and indigenous) where in 33.3 per cent of the households, women owned some chickens. This was followed by goats, where in 32.7 per cent of the households, women owned some goats (Figure 3.2).

Across the three countries, women owned cattle in 25.1 per cent of the sampled households. The lowest percentage was in Tanzania where women owned cattle in only 7.4 per cent of the households and the highest was in Mozambique where women owned cattle in 41 per cent of the households. Women ownership of goats was higher overall than for cattle although women still owned goats in less than 35 per cent of the households.

In Tanzania, the proportion of households where women owned goats was lower than the proportion of households where they owned cattle. In over 50 per cent of the households in Mozambique, women owned poultry. The high ownership of poultry and goats by women has been documented in other countries. For example in the Gambia, Jaitner *et al.* (2001) found that women owned 52 per cent of goats in livestock-keeping households while studies in Kenya and Uganda found 63 per cent and 23 per cent of chickens respectively were owned by women (Okitoi *et al.* 2007; Oluka *et al.* 2005).

Despite these high proportions of households where women owned different livestock species, the proportion of livestock that they own and the average numbers they own are much less that that owned by men (Figure 3.3 and Table 3.1).

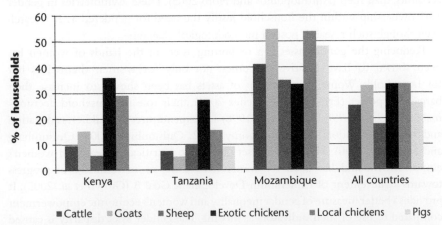

FIGURE 3.2 Percentage of households where women own different livestock species

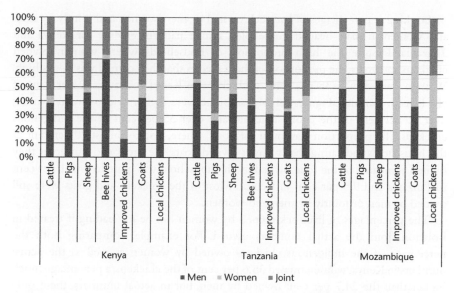

FIGURE 3.3 Proportion of livestock owned by men, women and jointly in male-headed households

TABLE 3.1 Average numbers of livestock owned by men, women and jointly in male-headed households

		Kenya			Tanzania			Mozambique		
	Statistics	Men	Women	Joint	Men	Women	Joint	Men	Women	Joint
Cattle	Mean	1.0	0.1	1.4	3.6	0.2	3.0	7.3	6.0	1.4
	SD	2.0	0.5	2.0	7.2	0.8	3.8	11.5	12.4	5.1
Pigs	Mean	2.8	0.0	3.5	1.1	0.2	2.8	5.3	3.1	0.4
	SD	3.8	0.0	6.7	2.5	0.8	4.9	6.9	5.6	1.6
Sheep	Mean	1.7	0.1	1.9	2.6	0.6	2.5	4.8	3.4	0.5
	SD	2.8	0.6	3.4	4.0	1.9	4.3	7.0	7.4	1.8
Exotic	Mean	11.5	32.1	44.1	65.1	42.7	99.4	0.0	369.3	5.7
chickens	SD	30.9	81.0	105.9	156.1	91.0	178.2	0.0	639.7	4.9
Goats	Mean	2.5	0.6	2.8	2.9	0.2	5.5	5.0	5.7	2.7
	SD	16.9	1.6	3.6	6.3	0.8	7.0	8.6	10.4	7.4
Local	Mean	3.0	4.4	4.9	5.7	6.1	14.8	2.9	4.9	5.3
chickens	SD	17.4	14.9	8.6	18.8	24.7	26.8	7.9	8.4	9.0

Looking at percentages of livestock owned by men, women and jointly by men and women in male-headed households, similar patterns emerge for Kenya and Tanzania. Despite the high proportion of households where women owned chickens and goats, the proportion of these species that they owned was not as high. Women owned over 35 per cent of chickens in Kenya and over 20 per cent of chickens in Tanzania. Figures for Mozambique were higher, with women owning over 90 per cent of the exotic chickens and over 35 per cent of the indigenous chickens. This proportion was lower for goats, where, despite almost 20 per cent of the households indicating that women owned goats, the proportion of goats they owned compared to the total owned by the household was quite low (2.1 per cent in Tanzania and 9.5 per cent in Kenya). Most of the chickens and goats were still owned by men or jointly by men and women.

The percentages of livestock owned by women can be misleading if treated in isolation from the actual numbers owned. For example, comparing both the percentage of the indigenous chickens owned by women as well as the actual numbers in Kenya, women owned 35.6 per cent of the chickens, a percentage much higher than the 24.7 per cent owned by men, but in actual numbers, these only comprised an average of 4.4 birds. A similar trend is observed in Tanzania where women owned 23.1 per cent of the indigenous chickens at an average of 6.1 birds. For exotic chickens, women owned three times more than men in Kenya, more than one and a half times in Tanzania, and all the exotic chickens in Mozambique. In both Kenya and Tanzania, joint ownership of livestock was more common than in Mozambique. In Kenya, for example, joint ownership accounted for over 50 per cent of the cattle, sheep, goats and exotic chickens owned by the household. For the same species, joint ownership accounted for less than 10 per cent in Mozambique.

The patterns of livestock ownership differ substantially across the three countries. In Mozambique, women owned 40.7 per cent of the cattle, compared to 5.2 per cent and 2.7 per cent in Kenya and Tanzania respectively and owned on average 6 head compared to 0.1 and 0.2 head of cattle for Kenya and Tanzania respectively. Other studies from Southern Africa have reported similar ownership patterns for livestock as in Mozambique. In Zimbabwe Chawatama *et al.* (2005) reported from their study in three districts that women owned on average 6.1, 4.5 and 5.2 head of cattle in Chikomba, Kadoma and Matobo districts respectively. In Botswana, Oladele and Monkhei (2008) found that women owned 25 per cent of the cattle and 81 per cent of goats.

Looking at the gender disparity in ownership of livestock, in Kenya, men owned 10 times more cattle than women, while in Tanzania, men owned 18 times more cattle than women. Mozambique had the lowest gender disparity in cattle ownership, with women owning 0.8 head for every 1 head of cattle that men owned. Goat ownership exhibited similar trends in Kenya and Tanzania. In Kenya, for every 1 goat owned by women, men owned 4 goats while in Tanzania, for every 1 goat owned by women, men owned 14 goats. Ownership of local chickens was higher for women than men in all countries. However, ownership of improved chickens was higher for men in Tanzania, where men owned one and a half times more improved chickens than women. The differences in actual numbers owned by men and women across all species other than chickens were, however, not significant in Mozambique. Despite the fact that women were more likely to own chickens and goats, in all countries, they did not own higher numbers of goats than men, and in fact in Tanzania, men owned a significantly higher number of goats than women.

Decision-making on women-owned livestock

The concept of ownership cannot be taken in isolation from decision-making. For women-owned assets, it is essential to establish whether they can sell, give out and slaughter, and whether they can make the decisions independently or have to consult other members of the household, especially their husbands. According to Deere and Doss (2006) these questions provide a more nuanced understanding of women's livestock ownership and what rights and responsibilities are attached to the livestock assets that women own. We use data from Kenya and Tanzania to explore this. In Kenya, for women-owned livestock, less than half of women could sell their local and improved chickens without consulting their husbands (37.5 per cent and 34.8 per cent respectively for local and indigenous chickens). The proportion was even lower for larger livestock, whereby only 8.8 per cent, 13.8 per cent and 10.0 per cent of women could sell their dairy cattle, sheep and goats respectively without consulting their husbands. In Kenya, 43.1 per cent, 36.2 per cent and 30 per cent of women indicated that their husbands could sell dairy cattle, sheep, goats and pigs respectively that they owned without having to consult them (see Figure 3.4).

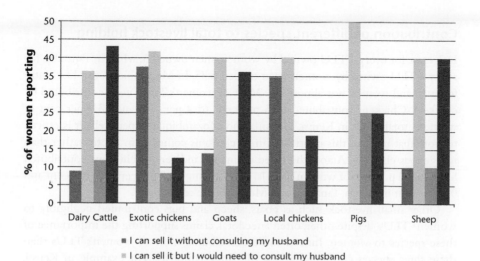

FIGURE 3.4 Decision-making on sale of women-owned livestock in Kenya

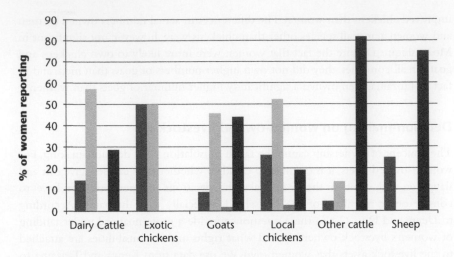

I can sell it without consulting my husband
I can sell it but I would need to consult my husband
My husband is the only one who can sell it. He does not have to consult me
My husband can sell it but he would have to consult me

FIGURE 3.5 Decision-making on sale of women-owned livestock in Tanzania

In Tanzania, half of the women interviewed could sell the exotic chickens that they owned without having to consult their husbands while the other 50 per cent had to consult their husbands before they could sell. For cattle and sheep, the husbands could sell, but they would have to consult the wives as owners (Figure 3.5).

Contribution of different species to total livestock holding

In Kenya, for every 1 TLU owned by women, men owned 5, while in Tanzania, for every 1 TLU owned by women, men owned 6.35. In Mozambique, the TLU ratio for men and women was almost 1:1. Most of women's TLUs were contributed by chickens. Chickens contributed 86.5 per cent, 61.4 per cent and 33.9 per cent of women's total TLUs in Kenya, Tanzania and Mozambique respectively. It is crucial to note that most of this contribution of chickens comes from exotic rather than indigenous chickens. Across the three countries, indigenous chickens contributed less than 10 per cent of women's TLUs, contributing 10.5 per cent, 7.7 per cent and 0.4 per cent in Kenya, Tanzania and Mozambique respectively.

Other small livestock such as goats, sheep and pigs contributed negligibly to women's TLUs, despite other, often anecdotal, claims supporting the importance of these species to women. Indeed cattle contributed more to women's TLUs than these three species combined across the three countries. For example in Kenya, cattle contributed 30.9 per cent of women's TLUs, whereas goats and sheep contributed only 13.2 and 3.1 per cent respectively. Similar patterns were observed in Tanzania and Mozambique. In Tanzania, cattle contributed 23.3 per cent of the

TABLE 3.2 TLUs held by men, women and jointly by men and women in Kenya, Tanzania and Mozambique

Kenya

	TLU (men-owned livestock)	TLU (women-owned livestock)	TLU (Joint men- and women-owned livestock)	t-values (men-owned, women-owned)
Cattle	0.97 (46.1%)	0.13 (30.9%)	1.42 (46.1%)	4.699★★★
Pigs	0.57 (26.9%)	0.00 (0.0%)	0.70 (22.7%)	1.818
Sheep	0.17 (8.2%)	0.01 (3.1%)	0.19 (6.1%)	3.942★★★
Exotic chickens	0.12 (5.5%)	0.32 (76.3%)	0.44 (14.3%)	1.130
Goats	0.25 (11.8%)	0.06 (13.2%)	0.28 (9.2%)	1.012
Local chickens	0.03 (1.4%)	0.04 (10.5%)	0.05 (1.6%)	0.680
Total mean	2.10	0.42	0.77	

Tanzania

	TLU (men-owned livestock)	TLU (women-owned livestock)	TLU (Joint men- and women-owned livestock)	t-values (men-owned, women-owned)
Cattle	3.61 (71.0%)	0.19 (23.3%)	2.98 (54.3%)	3.483★★★
Pigs	0.22 (4.3%)	0.04 (5.5%)	0.56 (10.3%)	2.396★★★
Sheep	0.26 (5.1%)	0.06 (7.5%)	0.25 (4.6%)	1.309
Exotic chickens	0.65 (12.8%)	0.43 (53.7%)	0.99 (18.1%)	0.377
Goats	0.29 (5.6%)	0.02 (2.2%)	0.55 (10.1%)	4.109★★★
Local chickens	0.06 (1.1%)	0.06 (7.7%)	0.15 (2.7%)	0.199
Total mean	5.08	0.80	5.49	

Mozambique

	TLU (men-owned livestock)	TLU (women-owned livestock)	TLU (Joint men- and women-owned livestock)	t-values (men-owned, women-owned)
Cattle	7.15 (78.2%)	5.83 (52.9%)	1.36 (73.5%)	0.931
Pigs	1.01 (11.1%)	0.58 (5.3%)	0.08 (4.3%)	1.430
Sheep	0.47 (5.1%)	0.32 (2.9%)	0.04 (2.3%)	0.844★
Exotic chickens	0.00 (0.00%)	3.69 (33.5%)	0.06 (3.1%)	1.000
Goats	0.49 (5.4%)	0.56 (5.1%)	0.26 (14.1%)	0.797
Local chickens	0.03 (0.3%)	0.05 (0.4%)	0.05 (2.8%)	2.250★
Total mean	9.14	11.04	1.86	

Numbers in brackets are percentage contribution of the species to TLUs.
★★★, ★★, ★ significant at 1%, 5% and 10% respectively.

women-owned TLUs to goats' 2.2 per cent and sheep's 7.5 per cent. In Mozambique, sheep and goats contributed 2.9 per cent and 5.1 per cent of the women-owned TLUs, while cattle contributed 52.9 per cent. While some of the small stock does not contribute significantly to women's total TLUs, women often have more decision-making authority over these than they do over large animals such as cattle.

Means of acquisition of livestock by women

In order to help women secure, build and safeguard their assets, a better understanding of how households accumulate livestock can inform the design and implementation of development interventions (Kristjanson *et al.* 2010). In Kenya, the main means of livestock acquisition for women was through purchases (50.4 per cent) and livestock born into the herd (28.5 per cent). Over 50 per cent of female-owned cattle, sheep and exotic chickens were purchased, while 31.9 per cent and 48.6 per cent of the goats and local chickens were purchased (see Figure 3.6). Grants from non-governmental organizations (NGOs) and other externally funded projects were an important source of goats and exotic chickens for women, with 25.5 per cent of the goats and 30.4 per cent of the exotic chickens owned by women coming from grants. Inheritance and group purchase were not common sources of livestock for women, with only 1.7 per cent and 3.1 per cent of women-owned livestock being acquired through these two means.

As in Kenya, most of the women-owned livestock in Tanzania was either purchased (52.6 per cent) or born into the herd or flock (37.6 per cent). Over 50 per cent of the cattle, goats and pigs were purchased (see Figure 3.7). All the women-owned pigs and exotic chickens were purchased. Unlike in Kenya, where inheritance of livestock by women was not common, women in Tanzania inherited

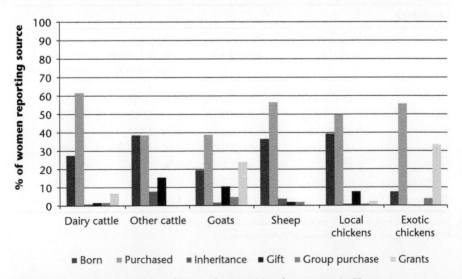

FIGURE 3.6 Common means of livestock acquisition by women in Kenya

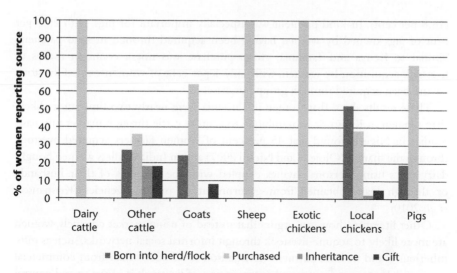

FIGURE 3.7 Common means of livestock acquisition by women in Tanzania

10.5 per cent of the cattle they owned. Gifting was also a common source of livestock with 5.3 per cent of all women-owned livestock having been received as a gift, the most common species acquired in this way being goats (8.3 per cent).

Figure 3.8 shows purchase of livestock was highest in Mozambique, with women purchasing 72.7 per cent of the livestock they own. Most of the goats were acquired through purchase (81.6 per cent) as were the local chickens (72.3 per cent). There was also more diversity of livestock sources in Mozambique compared to Kenya and Tanzania. For example, women acquired cattle from purchase (56.3 per cent), inheritance (6.3 per cent), as a gift (18.8 per cent) and as in-kind payment

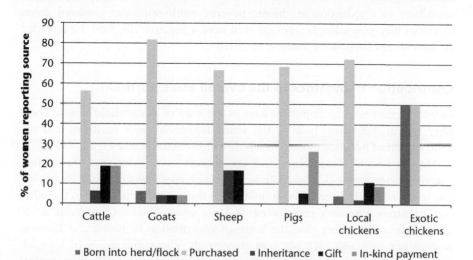

FIGURE 3.8 Common means of livestock acquisition by women in Mozambique

(18.8 per cent). In-kind payment was especially important for pigs, with 26.3 per cent of pigs owned by women having been acquired through in-kind payment. Similar to Kenya and Tanzania, group purchase was not a common means of livestock acquisition by women, with only 1 per cent of women-owned livestock having been acquired through group purchase.

These results show that women in Tanzania were nearly two times more likely than women in Kenya or Mozambique to acquire cattle through market purchase. Studies in Nigeria found that 45 per cent of women farmers acquired livestock through the market (Olojede and Njoku 2007) and in India landless women bought dairy cows using personal savings coupled with the earnings of their husbands, or through loans obtained from government or private agencies (Kristjanson et al. 2010).

Other literature, however, argues that instead of using market channels, women are more likely to acquire livestock through informal social networks, such as gifts, inheritance and in-kind payment (Kristjanson et al. 2010) than from commercial markets. A key constraint to market acquisition of livestock is access to and control over capital. Findings from a study in Zimbabwe support this and show that 60 per cent of women lack the capital to purchase livestock because men control cash incomes generated from crop and livestock sales (Chawatama et al. 2005). These and other studies (Rubin et al. 2010; Todd 1998) recommend the provision of micro-credit as one approach to reduce women's limited access to cash and enable their purchase of livestock. A less common form of livestock acquisition by women from the study was through grants, which were recorded only by women in Kenya. NGO grants are indicative of livestock development and redistributive programs. They are common in Kenya in the form of restocking, breed improvement and nutrition interventions, especially for livestock such as exotic chickens, dairy cattle and goats. While livestock grants can build up the assets of poor people and contribute to a reduction in chronic poverty, overlooking the gendered access dynamics may jeopardize benefits, or even have a negative effect on the intended women beneficiaries (Kristjanson et al. 2010).

Contribution of livestock to the overall asset portfolio

Livestock constitutes an important asset in the suite of a household's assets. There were critical differences both in the actual household asset portfolios and the contribution of livestock to these across the three countries (see Figure 3.9).

Households in Mozambique had the highest asset index and the highest contribution by livestock to assets. Indeed, the high asset index in Mozambique was due to the relatively large livestock holdings in this country compared to the other two countries. Livestock contributed 84.7 per cent of the total movable assets, which was much higher than the livestock contribution in Kenya and Tanzania at 51.9 per cent and 59.1 per cent respectively. These differences in livestock assets could be explained by the differences in the livestock production systems in the three countries. While in Kenya the study was mainly done in the mixed

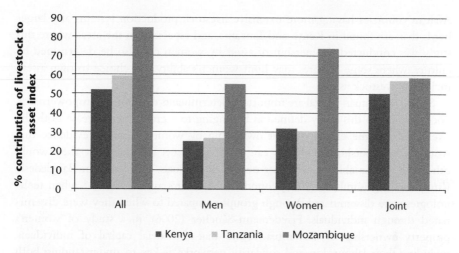

FIGURE 3.9 Contribution of livestock to household, men's and women's overall asset portfolio

crop-livestock systems, the Mozambique sites were drier; the environment more suitable for livestock production and households therefore had higher numbers of livestock than at the Kenya and Tanzania sites.

In terms of ownership patterns, most of the assets in Kenya (more than 50 per cent) and Tanzania (slightly below 50 per cent) were jointly owned, while in Mozambique most of the assets were owned either by men or by women, with only about 11 per cent owned jointly. In all three countries, livestock were much more important to women's asset portfolio than men's. For example in Mozambique, livestock contributed to 55 per cent of men's assets and 73.8 per cent of women's assets. In Kenya and Tanzania, while the livestock contribution was much lower than for Mozambique, livestock contributed to one-third of women's total movable asset portfolio (31.8 per cent and 30.4 per cent respectively). It is important to note that ownership and rights over livestock are quite complex and that women can derive benefits from livestock irrespective of whether they own them or not. For example, a woman may have the rights to obtain milk from certain animals, even if she does not formally own these animals.

What determines women's ownership of livestock?

Livestock ownership by women can be determined by various factors. There are variations across countries in both the proportion of households where women owned livestock and the numbers of livestock that women owned. While there is evidence of higher women ownership of livestock in the Southern African countries, with evidence from Botswana (Oladele and Monkhei 2008) and Zimbabwe (Chawatama *et al.* 2005), and Mozambique (this chapter), we did not find any documented evidence of the reasons for this variation. It could be explained, however, by the differences in production systems, with the Southern

African countries having more extensive livestock production systems and larger herds than are found in Kenya and Tanzania, and especially in the areas where this study was conducted. Ownership of assets by women can also be determined by culture, where some cultures may limit women's ability and choice to own assets, including livestock assets.

Social and human capital are important determinants of women's empowerment. Social capital, in this study defined as belonging to a group, can increase both the likelihood and the extent of asset ownership by women. There is evidence that women's membership in groups can facilitate access to assets that they would otherwise not be able to access or own as individuals. In a study in Bangladesh, Kumar and Quisumbing (2010) found that women's assets increased when technologies were disseminated through groups compared to when they were disseminated through individuals. Friedemann-Sánchez (2006), in a study of women's property ownership in Colombia, found that the social capital of individuals, including their labour, kin and solidarity networks, is key to understanding both property acquisition and intra-household bargaining processes.

Human capital can take many forms, including labour available to households, health and education. Education plays a major role not only for individuals' opportunities in society, but also for the productive capacity and well-being of a household. We hypothesized that women's education would have an impact on their asset ownership through various mechanisms: first, that more educated women would have multiple opportunities for income and asset accumulation due to the opportunities accorded by higher education and, second, that women with higher education and therefore higher levels of human capital are able to bargain for ownership of household resources. The International Fund for Agriculture Development (IFAD 2001) sees human assets, including education, as having two types of values: an *intrinsic value* in raising capabilities, which can have psychological benefits in terms of self-esteem or happiness (which do not necessarily translate into instrumental value); and *instrumental value* in raising productivity and income, which further enhances the intrinsic value or benefits. It is expected that women's education can have an instrumental value, which enables them to use it to further accumulate other assets such as livestock. There is ample evidence that, in most countries, women have less education than their male counterparts (World Bank 2001) although the gender education gap is narrowing.

The ownership of other assets by women was expected to have an influence on women's ownership of livestock. This hypothesis was built upon the asset ladder, where women were expected to own small assets, such as domestic assets, farm implements and, with accumulation of these assets and a strengthening of their bargaining and voice within the household, acquire larger assets such as livestock.

Across countries, there are variations in the ownership of livestock by women. We analysed the factors that influence the ability of women to own livestock in two key stages, in the first stage we used a probit analysis to determine what influences whether women own livestock and in the second stage we used a linear regression model to determine the extent of ownership of livestock by women using the women-owned TLUs as a dependent variable.

TABLE 3.3 A probit analysis of factors influencing women's ownership of livestock in Kenya, Tanzania and Mozambique

Women own livestock (1 = yes, 0 = no)	Coefficient	T		
Age of spouse	0.002	0.6	0.098	2.74★★★
Other assets index by women	0.009	2.24★★	0.001	0.13
Belong to a group (1 = yes)	−0.001	−0.1	0.111	3.63★★★
Primary education (1 = yes)	−0.006	−0.04	−0.153	−0.08
Above primary education (1 = yes)	0.095	0.44	−0.678	−0.47
Kenya	−1.519	−8.89★★★	−8.044	−5.34★★★
Tanzania	−1.658	−8.96★★★	−6.907	−3.8★★★
Constant	0.791	4.27★★★		
Number of observations	469.00		229	
Design df	468.00		228	
F(7, 462)	19.47		5.44	
Prob > F	0.00		0.00	

★★★, ★★, ★ significant at 1%, 5% and 10% respectively.

Table 3.3 shows that the probability of women owning livestock increased with women's ownership of other domestic assets and the factors that influence the value of women's TLUs. The probability of women owning livestock was also higher in Mozambique, than in both Kenya and Tanzania, as expected given the higher percentages of households where women owned livestock in Mozambique.

Belonging to a group or having primary or higher education did not influence the probability of women owning livestock as was expected. It did, however, influence how many TLUs women had, with women who belonged to a group having more TLUs than those who did not belong to a group. The finding that belonging to a group increased women's livestock asset index by 0.11 points is supported by the finding by Kumar and Quisumbing (2010), in their analysis of impacts of collective and individual approaches for technology dissemination on gender asset disparities in Bangladesh, that in the cases where group approaches were used, women increased their assets more than in cases where individual approaches were used. While ownership of other assets increased the probability that women would own livestock, it did not influence the numbers of livestock owned by women. Women from Mozambique, as expected, owned more livestock than women in Kenya and Tanzania.

Conclusions

Patterns of livestock ownership varied across the three countries. The results of the analysis suggest that the analysis of livestock ownership by women should use multiple methods and look at different dimensions of ownership. Chickens were the species most commonly owned by women. Poultry are important livestock for women. In the three countries, poultry contributed the highest proportion of the

TLUs owned by women. Despite evidence on the role of small ruminants as an asset for women, goats contributed negligibly to women's total TLUs. Cattle, on the other hand, contributed to a significant proportion of women's TLUs. For research and development programs working to increase the value of women's assets through livestock, cattle are still a very attractive option. Care must be taken, however, to ensure that women do not lose ownership and control of the cattle, as evidence shows that larger animals are more likely to be controlled by men than by women. Also, given the variation in importance of different livestock for women, a species focus should be carefully guided by this, rather than by a general assumption that small livestock are the most important for women, or that women are more likely to own these species. Further research that looks at species ownership alongside benefits that women get from these species would be useful, as women may own fewer of a particular species but derive more benefits from that species than another species where they own more.

The evidence suggests that even when the proportions of households where women own livestock are high, women still own fewer livestock compared to men. Increasing access to livestock by women should focus not only on having more women own livestock, but also on ensuring that the gender gap in livestock ownership is reduced. It is also not enough to rely on grants and group purchases for increasing women's livestock ownership. Increasing women's access to credit and designing innovative mechanisms such as livestock leasing schemes, where women access livestock and repay through product sales, should be explored and scaled out where these are found to work.

The concept of asset ownership is complex and can differ depending on the cultural context as well as the production system. There is a need to use participatory approaches to understand what ownership means for both men and women before collecting data on asset ownership. Ownership may imply legal ownership, where the person legally has a title to an asset or property. This is mainly applicable to assets such as land. For livestock ownership, however, there is no legal title or document to show ownership. Women mainly said they owned livestock because they had purchased the animals using their own generated income, had received the livestock individually through grants from NGOs, or had purchased them with income earned from other activities. Contrary to other evidence, purchase was still the most common means of livestock acquisition by women. This type of ownership, however, did not mean that women always had decision-making authority, or control over these livestock.

Compared to other household assets, livestock are an important asset for women and contribute to a significant proportion of women–owned assets. There are factors that increase the probability that women will own livestock. The role of groups in helping women accrue assets cannot be over-emphasized, despite the low numbers of women acquiring livestock through group purchases. Social capital may serve different functions, which include helping women save money that can be used to purchase livestock or increase access to credit, other financial resources and output markets, all of which can play a role in helping women accumulate assets.

References

Agarwal, B. (1998) Widows vs. daughters or widows as daughters? Property, land and economic security in rural India. In M. A. Chen (ed.) *Widows in India: Social Neglect and Public Action*. New Delhi: Sage, pp. 124–169.

Agarwal, B. (2002) *Bargaining and Legal Change: Toward Gender Equality in India's Inheritance Laws.* Working Paper. Brighton, UK: Institute for Development Studies.

Allendorf, K. (2007) Do women's land rights promote empowerment and child health in Nepal? *World Development* 35(11): 1975–1988.

Antonopoulos, R. and Floro, M. (2005) Asset ownership among poor households in Bangkok. *Journal of Income Distribution*, Review of Income and Wealth.

Banerjee, A. V. and Duflo, E. (2003) Inequality and growth: what can the data say? *Journal of Economic Growth* 8(3): 267–299.

Barham, B., Carter, M. R. and Sigelko, W. (1995) Agro-export production and peasant land access: examining the dynamic between adoption and accumulation. *Journal of Development Economics* 46(1): 85–107.

Birdsall, N. and Londono, J. L. (1997) Asset inequality matters: an assessment of the World Bank's approach to poverty reduction. *American Economic Review* 87(2): 32–37.

Carney, D. with Drinkwater, M., Rusinow, T., Neefjes, K., Wanmali, S. and Singh, N. (1999) *Livelihoods Approaches Compared. A Brief Comparison of the Livelihoods Approaches for the UK Department for International Development (DfID), CARE, Oxfam, and the United Nations Development Programme (UNDP).* London: DfID.

Chawatama, S., Mutisi, C. and Mupawaenda, A. C. (2005) The socio-economic status of smallholder livestock production in Zimbabwe: a diagnostic study. *Livestock Research for Rural Development* 17, Article #143. Available at: http://www.lrrd.org/lrrd17/12/chaw17143.htm (accessed 29 June 2011).

Deere, C. D. and Doss, C. R. (2006) The gender asset gap: what do we know and why does it matter? *Feminist Economics* 12(1–2): 1–50.

DfID (1997) *The UK White Paper on International Development and Beyond.* London: DfID.

Doss, C., Grown, C. and Deere, C. D. (2007) *Gender and Asset Ownership: A Guide to Collecting Individual Level Data.* Policy Research Working Paper 4704. Washington, DC: World Bank.

Doss, C., Truong, M., Nabanoga, G. and Namaalwa, J. (2011) *Women, Marriage and Asset Inheritance in Uganda.* Chronic Poverty Research Centre Working Paper No. 184. New Haven, CT: Yale University.

Duraisamy, P. (1992) Gender, intrafamily allocations of resources and child schooling in South India. *Economic Growth Center, Discussion Paper No. 667.* New Haven, CT: Yale University.

Ford Foundation (2004) *Building Assets to Reduce Poverty and Injustice.* New York: Ford Foundation.

Friedemann-Sánchez, G. (2006) Assets in intrahousehold bargaining among women workers in Colombia's cut-flower industry. *Feminist Economics* 12(1–2): 247–69

Grown, C., Gupta, G. R. and Kes, A. (2005) *Taking Action: Achieving Gender Equality and the Millennium Development Goals.* London: Earthscan.

IFAD (2001) *Rural Poverty Report.* Rome: IFAD.

Jaitner, J., Sowe, J., Secka-Njie, E. and Dempfle, L. (2001). Ownership pattern and management practices of small ruminants in The Gambia – implications for a breeding programme. *Small Ruminant Research* 40(2): 101–108.

Kristjanson, P., Waters-Bayer, A., Johnson, N., Tipilda, A., Njuki, J., Baltenweck, I. et al. (2010) *Livestock and Women's Livelihoods: A Review of the Recent Evidence.* ILRI Discussion Paper No. 20. Nairobi: ILRI.

Kumar, N. and Quisumbing, A. (2010) *Does Social Capital Build Women's Assets? The Long-term Impacts of Group-based and Individual Dissemination of Agricultural Technology in Bangladesh.* CAPRi Working Paper No. 97. Washington, DC: IFPRI.

Lundberg, S. and Pollak, R. (1996) Bargaining and distribution in marriage. *Journal of Economic Perspectives* 10(4): 139–158.

Marchant, M. A. (1997). Bargaining models for farm household decision making: discussion. *American Journal of Agricultural Economics* 79(2): 602–604.

Mason, K. (1998) Wives' economic decision-making power in the family: five Asian countries. In K. Mason (ed.) *The Changing Family in Comparative Perspective: Asia and the United States.* Honolulu: East-West Center.

Meinzen-Dick, R., Johnson, N., Quisumbing, A., Njuki, J., Behrman, J., Rubin, D. *et al.* (2011). *Gender, Assets, and Agricultural Development Programs: A Conceptual Framework.* Collective Action and Property Rights (CAPRi) Working Paper No. 99. Washington, DC: IFPRI.

Okitoi, L. O., Ondwasy, H. O., Obali, M. P. and Murekefu, F. (2007) Gender issues in poultry production in rural households of western Kenya. *Livestock Research for Rural Development* 19, Article #17. Available at: http://www.lrrd.org/lrrd19/2/okit19017.htm (accessed 29 June 2011).

Oladele, O. and Monkhei, M. (2008) Gender ownership patterns of livestock in Botswana. *Livestock Research for Rural Development* 20(10).

Olojede, J. C. and Njoku, M. E. (2007) Involvement of women in livestock production: a means of reducing hunger and malnutrition in Ikwuano local government area of Abia State, Nigeria. *Agricultural Journal* 2: 231–235.

Oluka, J., Owoyesigire, B., Esenu, B. and Sssewannyana, E. (2005). Small stock and women in livestock production in the Teso Farming System region of Uganda. *Small Stock in Development* 151.

Quisumbing, A. R. (2003). What have we learned from research on intra-household allocation? In A. Quisumbing (ed.) *Household Decisions, Gender, and Development: A Synthesis of Recent Research.* Washington, DC: IFPRI.

Quisumbing, A. and de la Brière, B. (2000) *Women's Assets and Intrahousehold Allocation in Rural Bangladesh.* Washington, DC: IFPRI.

Quisumbing, A. R. and Maluccio, J. A. (2000) *Intrahousehold Allocation and Gender Relations: New Empirical Evidence from Four Developing Countries.* Washington, DC: International Food Policy Research Institute.

Rubin, D., Tezera, S. and Caldwell, L. (2010) *A Calf, a House, a Business of One's Own: Microcredit, Asset Accumulation, and Economic Empowerment in GL CRSP Projects in Ethiopia and Ghana.* Washington, DC: Global Livestock Collaborative Research Support Program.

Scoones, I. (2009) Livelihoods perspectives and rural development. *Journal of Peasant Studies* 36(1): 171–196.

Todd, H. (1998) Women climbing out of poverty through credit: or, what do cows have to do with it? *Livestock Research for Rural Development* 10(3): 1–3.

Torkelsson, A. and Tassew, B. (2008) Quantifying women's and men's rural resource portfolios: empirical evidence from Western Shoa, Ethiopia. *European Journal of Development Research* 20(3): 462–481.

World Bank (2001) *Engendering Development.* Washington, DC: World Bank.

4
GENDERED PARTICIPATION IN LIVESTOCK MARKETS

*Elizabeth Waithanji, Jemimah Njuki and
Bagalwa Nabintu*

Background

The rapid change experienced in livestock markets in the last few decades has been attributed to the increasing demand for livestock products in both developed and developing countries owing to increases in incomes among some urban populations. Between the 1970s and 1990s for example, annual per capita meat consumption more than tripled in developing countries. Milk consumption also increased, but to a lesser extent than meat consumption, in both economies (Delgado 2003). This trend appears to continue as demonstrated by Kristensen *et al.* (2004), who suggest deliberate targeting of smallholder farmers for production of the needed animal-based food. The increase in demand for livestock and their products offers an opportunity for growth of livestock markets and participation in these markets by smallholder farmers. Participation in markets by smallholders is determined by numerous costs and benefits, such as transaction costs, which may or may not be compensated for by high revenues; prices; turnover; uncertainty; cooperation and collective initiatives; and labour and capital investment (Verhaegen and Van Huylenbroeck 2001). These and other costs and benefits vary with gender. This chapter looks at the gendered differences in participation by smallholder farmers in livestock markets and explains the differences using some feminist and other theories.

Although some empirical evidence exists in terms of women's participation in crop and labour markets (Zaal 1999; Njuki *et al.* 2011a) and in extensive pastoral and agro-pastoral livestock production systems (Fratkin and Smith 1995; Nunow 2000), much less survey-based research has been done on patterns of market participation by men and women smallholder livestock farmers across livestock and their products, and what determines these patterns. This empirical gap is particularly important because substantial controversy appears to be developing around two issues, namely: what factors influence women's participation in product markets and how does their participation change as markets become more formalized? And are women more likely to participate in the marketing of small livestock and livestock products and mainly in informal markets? A lot of the evidence so far has been anecdotal and, where data exist, they have been on a limited number of products without comparative analysis across livestock species, products and types of markets. The research reported in this chapter undertakes to establish the types of markets and commodities (livestock species and products) in which women and men are involved.

Cross-livestock product and market comparisons will enable the identification of opportunities where women are likely to benefit from market participation and to develop some of the strategies that could be adopted to increase benefits to women from market participation. This chapter uses both qualitative data from the focus group discussions as well as quantitative data. The data is used to explain gendered patterns of participation for different livestock and products in different markets. While women's participation in livestock markets is an important way to improve the welfare of women and their families, it is also important that women are able to make decisions about which products and animals are sold and what is done with the proceeds of the sale, otherwise, participation alone may not benefit women (Kristjanson *et al.* 2010).

Livestock production and marketing systems

As agriculture and livestock production become more commercialized, women smallholder farmers may not be able to compete with and benefit like men smallholder farmers because women have a lower access to resources, including capital, than men, and they experience other social barriers unknown to men. In his work in Guatemala, Swetnam (1988) demonstrated a market-based sexual inequality, whereby most women sold goods carried high risk, the lowest profits and the least potential for amassing wealth. Among the Fulani of Nigeria and the Omduruman of Sudan, men seemed to become attracted by the increasing monetary importance of even traditionally women-controlled livestock products like milk and hides (Fratkin and Smith 1995), often reducing women's role to that of mere labourers (Nunow 2000). In most traditional pastoral production systems, women's priority is children's nutrition while that of men is herd growth. Traditionally, pastoral women determine what proportion of milk is to go to the children and calves, and therefore the balance between household food security and herd growth (Nunow 2000). Much of this division in gender roles is affected by the commercialization of livestock production.

Women also lack secure rights to production resources including land, labour and capital (Kabeer 2001; Moser 2006), have a lower human capital (Morrison and Jutting 2005) and are, therefore, less likely to be served by formal financial institutions than men. These constraints are in addition to the general constraints of high transaction costs that emanate from the lengthy channels of trade necessitated by long distances to markets, search for market outlets, transport to and from markets, lack of quality certification, disorganized brokers and agents, inability to pool products in order to benefit from economies of scale, and inter-seasonal and inter-regional variation of production (Fafchamps and Gabre-Madhin 2001).

Owing to their nature, livestock and livestock products go through different stages of the value chain. Women's participation at each of the levels of the value chain varies due to different factors, including their skills and capacities, access to capital, constraints on mobility and their ability to organize. Anecdotal evidence and some preliminary research work on livestock value chains (Njuki *et al.* 2011a) indicate that in a livestock value chain, the men:women ratio, in terms of representation and control, increases as the household wealth increases and as the value of milk increases. Often too, the market value of most agricultural commodities increases as the market location moves further away from the point of production due to the added costs of transporting the commodity.

Livestock and livestock product markets in Kenya, Tanzania and Mozambique some of which are considered in this chapter include dairy (cattle and goats), poultry (meat and eggs of predominantly exotic and indigenous chickens) and other red meats (beef, sheep and goats, and pork). Markets of these commodities are at different stages of development while some are better understood than others. For example, in Kenya, the milk market has been well studied. Smallholder dairy farmers produce up to 56 per cent of all milk produced in the country and market about 70 per cent of the milk they produce (Peeler and Omore 1997). Of the milk produced by a smallholder dairy farmer, 36 per cent is consumed at home by calves and household members and the remaining 64 per cent is marketed raw as surplus (Omore *et al.* 1999). Of the marketed surplus, 55 per cent is sold raw to individual and institution consumers, 38 per cent to marketing cooperatives and middle persons who then sell it raw to urban consumers and processors, and only 7 per cent is sold directly to processors (Omore *et al.* 1999).

In Mozambique live cattle and goat marketing is quite prominent. While little information exists on the volume of these sales, data from neighbouring Zimbabwe show that sales are often in informal markets, with many sales being at farm gate, where often farmers do not have any comparative prices. A study by ICRISAT showed that households with more than 20 goats sold only 13 per cent of their flock while those with small flocks sold as much as 36 per cent of their animals each year (van Rooyen and Homann n.d.). A typical goat value chain is shown in Figure 4.1.

Most of the research on women's roles in livestock marketing has been done in pastoralist areas and intensive systems. For example, a USAID (US Agency for International Development) project in the Mandera triangle (covering Kenya, Somalia and Ethiopia) documents women's participation in milk, sheep and goat markets. Women sold milk and butter to traders, restaurant owners and families in nearby towns. The amount of milk and milk products sold varied based on men's decisions on how many animals to keep close to home and towns when they migrated with animals in search of pasture (USAID 2009). According to Ridgewell and Flintan (2007), trading in milk provides women with one of the few available

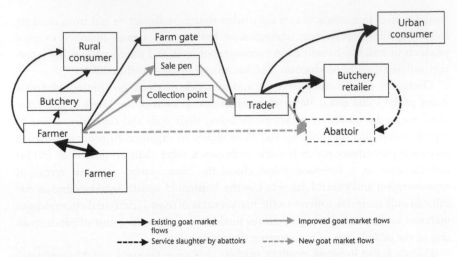

FIGURE 4.1 A typical goat value chain in Matabeleland, Zimbabwe (van Rooyen and Homann n.d.)

opportunities to control their own money. Although the movement with livestock is a constraint to women's organized marketing, the growth of settlements and urban centres has increased demand for milk and led to women being more organized to meet this demand (McPeak and Doss 2006; Ridgewell and Flintan 2007). The USAID study also documented in detail the means through which women transported milk (USAID 2009: 18):

> Milk marketing in the northern part of Kenya is exclusively the responsibility of women. On average, it took five hours to walk to the nearest town from the household in Chalbi and eight hours in Dukana. Milk production from the household herd averaged 4.5 litres per day in Chalbi and 3.5 litres per day in Dukana. The trips taken by wives to towns typically involve waking up pre-dawn, carrying some share of the milk collected the prior evening from the household herd in a small plastic or traditional woven container, and walking to town where they sell the milk themselves. They then use the income generated by these milk sales to make purchases before returning on foot to the household before night falls.

Unlike the cow milk market in northern Kenya, the camel milk trade is much more sophisticated, with women mainly acting as milk collectors, often based in mobile camps which follow seasonal partial transhumance (Nori *et al.* 2006; USAID 2009).

Women's participation in marketing of live animals, including cattle, sheep and goats, is much lower than their participation in the milk market. Evidence from Ethiopia and elsewhere (USAID 2009) suggests that many pastoral women play a significant role in the selling and buying of goats and sheep, but not cattle and camels. This is mainly because, for most women, access to livestock is by virtue of their relationships to men (husbands, fathers and sons) who control livestock (Ridgewell and Flintan 2007). It would seem that women tend to have far more rights to access and disposal of livestock products like milk, butter, cheese, ghee, hides and skins than they do over the live animal itself.

In both Tanzania and Mozambique, a gap exists in gender work in livestock value chains. In Tanzania, more than 90 per cent of the livestock population are of indigenous types, with characteristically low productivity, kept in the traditional sector, but well adapted to the existing harsh environment, including resistance to diseases (Njombe and Msanga 2009). Of the 18.8 million cattle found in the country about 560,000 are dairy cattle, which consist of Friesian, Jersey, Ayrshire breeds and their crosses to the East African Zebu. Seventy per cent of the milk is produced by Zebu cattle (Njombe and Msanga 2009). In Tanzania, research on livestock has focused more on the ecology and political economy of extensive livestock production systems, and on cattle, than other production systems and livestock species (for examples see Madox 1996; Fleisher 1998; Brockington 2001). Work on markets, too, is generalized for multiple commodities, with most discourses hinging on the fact that markets and the general infrastructure are poorly developed.

Even less is known about Mozambique livestock value chains. The limited research in livestock production reveals that the state of livestock production is low, and so is the state of consumption of livestock products. Mozambique ranks among the bottom 10 global meat-consuming countries and constitutes one of the bottom quartile countries of combined global meat and fish consumers (Speedy 2003). This is in spite of the continuing great global increase in production of livestock products, especially in poultry meat and eggs, milk and pig production (Speedy 2003). In contrast to the low livestock production, Mozambique counts as one of the few African countries that produce crop residues in excess of the amount that can be used by the existing livestock population (Kossila 1988). Mozambique, therefore, appears to be a promising livestock producer for the local and export markets. In a study on the cattle population in a part of southern Mozambique, 20 per cent of the cattle maintained were work oxen used for ploughing and transportation using small sledges. Farmers milked these draught cattle during the rainy season (Rocha et al. 1991).

Gender preferences for livestock and livestock products

Gendered preferences for livestock and livestock products were found to be determined by four main economic factors, namely: benefits from income; the security of owning the livestock as an asset; marketability of the livestock or product; and labour requirements for production and management of the livestock. Farmers' preferences could also be motivated by cultural and socio-economic incentives (Duvel and Stephanus 2000), which are less well explored than the more tangible economic ones (Jabbar et al. 1998).

Figure 4.2 shows the livestock species and product preferences of men and women. In Tanzania, the biggest differences in preference between men and women

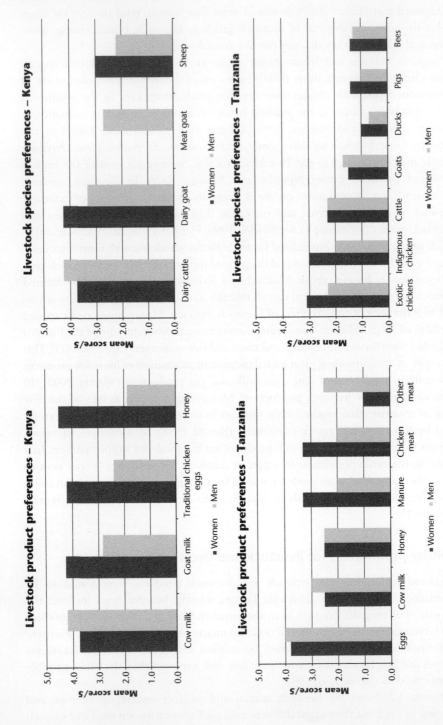

FIGURE 4.2 Livestock species and product preference of men and women in Tanzania and Kenya

were found in indigenous and exotic chickens, where women gave a much higher preference score than men did. Cattle, goats and bees received an almost equal preference score from both men and women. In terms of products, women gave a higher preference for manure than men did. All the other products (milk, eggs, honey) had an almost equal preference by men and women. The high preference for manure could be due to the importance manure plays in increasing crop production and productivity, and the important role that women play in crop production and in ensuring household food security.

In Kenya, there was a much more distinct gender difference in the preference for both the livestock species and the products. Women had a stronger preference for dairy goats, local chickens and dairy cows. The preference for chickens and dairy goats on the part of women compared to men could be due to the fact that both chickens and goats do not require the owner to be a land owner. Free range indigenous chickens often survive with minimal supplementation (Kitalyi 1998) while scavenging in backyards, while dairy goats can be zero grazed under the cut-and-carry (fodder) system. Women preferred indigenous chickens due to their low maintenance cost, disease resistance and marketability.

Dairy goats were preferred by women due to high kidding rates and the income earned from the sale of milk. The main difference between dairy goat milk and cow milk markets is that the goat milk market is predominantly informal, and although the milk is thought to have better nutritional quality than cow milk, the market remains relatively small and informal, and dominated by women.

Men had a higher preference score for dairy cattle and meat goats than did women. Men in Kenya found indigenous chickens to be undesirable because of their very low monetary value. These scores were similarly reflected in the products, where women had a much higher preference score for local eggs, honey and goat milk. Among the products, the main advantage cited for honey preference by both women and men was its medicinal value.

The main advantages of keeping dairy cattle cited by men in Kenya included the high value of cows and milk in monetary and nutrition (milk) terms. The disadvantages of keeping cattle, cited mainly by women in Kenya in reference to dairy cattle, included the high monetary cost of maintenance, too many labour demands and poor disease resistance. Men had a higher preference for cattle, both dairy and indigenous, than women. This could be explained by using Herskovits's (1924) concept of the East African cultural area. In the area, the cattle culture in which milk was used for subsistence and cattle for economic purposes was superimposed on the main agricultural culture. The fact that milk was for subsistence suggests that it was a women's product, and that cattle were kept for economic purposes suggests that cattle were men's commodities. This observation is supported by the following passage: "Cows in the north are sometimes tended by women and occasionally milked by them. This is never permitted in the south where only men must tend them" (Herskovits 1924: 50). As a vestige of this culture, therefore, cattle still remain predominantly, and sometimes exclusively, men's property in sub-Saharan Africa.

Eating roasted goat meat is a popular part of urban culture in both Tanzania and Kenya. The very well-developed urban goat-meat markets in Kenya may explain the exclusive preference for meat goats by men. Women in Kenya had less interest in meat goats because they saw no nutritional or monetary benefit from them. Women mentioned one of the disadvantages of the local goats predominantly used for the meat trade as destroying crops, as they are not reared using the cut-and-carry system that is common for dairy goats.

In Mozambique, women and men were asked to rank their livestock species preferences. Men preferred keeping cattle more than women, and women preferred keeping chickens more than men. Both women and men in Mozambique preferred raising sheep and goats almost equally. Owing to colonial and post-colonial land tenure systems and the more recent post-war land administration in Mozambique, women's initial control of everyday land management has been drastically eroded (Gengenbach 1998). As in Kenya and Tanzania, the preference for goats and chickens by women in Mozambique may be explained by their ability to keep these species on relatively small pieces of land. In order to own cattle, one needs to have some control over land. Because men own land and enjoy security of tenure, and because they are able to make decisions about the land as heads of families, they are able to keep land-dependent livestock like cattle in whatever numbers the available land can hold in all three countries.

As indicated in chapter 3, more women owned cattle in Mozambique than in Kenya and Tanzania. Compared to the East African countries, among communities in Mozambique, which constitutes the southern part of the eastern cultural area, cattle have a stronger cultural importance for men than in other more northerly areas (Tanzania and Kenya) because of the cultural influence of the Southern African communities that Herskovits (1924) termed the "Bushmen" and "Hottentots", whose main culture was cattle based. In these communities, men owned the cattle, but women milked them.

Patterns of market participation for different products

A quantitative analysis of the proportion of women and men selling different livestock products in markets in Kenya, Tanzania and Mozambique was conducted to help establish what markets existed for each product in each country, and who (men and women) sold in what market. Men and women were asked what livestock species and products they sold as individuals and jointly if married, and in which markets.

Several different markets that both men and women sold to were identified. They included farm gate to other farmers; farm gate to traders; delivery to traders; village market; city market; and specialized markets such as for honey, or collection centres and chilling plants for milk. It was expected that there would be price differences between these markets as well as differences in the costs of marketing.

Marketing patterns in Kenya

In Kenya, women sold 63.9 per cent of the total value of chickens sold while men only sold 6.5 per cent. Another 28.7 per cent of the total value of chickens sold by the households was sold jointly by men and women as shown in Figure 4.3. These results match with the preferences expressed by women during the focus group discussions and are in line with other results from Kenya. A study in Kajiado district, Kenya, found that women were the main sellers of chickens, and sold the birds to buy household provisions and feed and drugs for the remaining chickens, which were kept to be sold as a business (Muthiani *et al.* 2011). Women's participation in chicken and egg markets was higher than in other products. In another study in peri-urban areas of Kenya, Ngeno *et al.* (2011) found that in over 80 per cent of households, chickens were sold by women, and in over 95 per cent of households, eggs were also sold by women.

The most common market for women for chickens was farm gate, with 70 per cent of the total value of chickens sold by women being sold at farm gate to other farmers and traders, as shown in Figure 4.4. This is in contrast to men, who did not sell any chickens at farm gate to other farmers, although they did sell 29 per cent of the chickens they sold at farm gate to other traders. The most common chicken market for men was delivery to traders, where men sold 56 per cent of the total value of the chickens they sold. Women only delivered 22 per cent of their chicken sales to other traders.

Similar patterns were observed for sale of eggs and milk. Of the total value of eggs sold by households, women sold 89.1 per cent, while men sold only 8.9 per cent. Only 2 per cent of egg sales were done jointly by men and women. Of the eggs sold by women, close to 69 per cent were sold at farm gate to other farmers

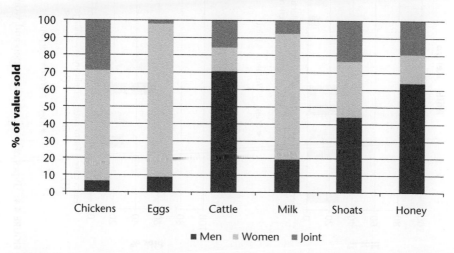

FIGURE 4.3 Percentage value of different products sold by men, women and jointly in Kenya

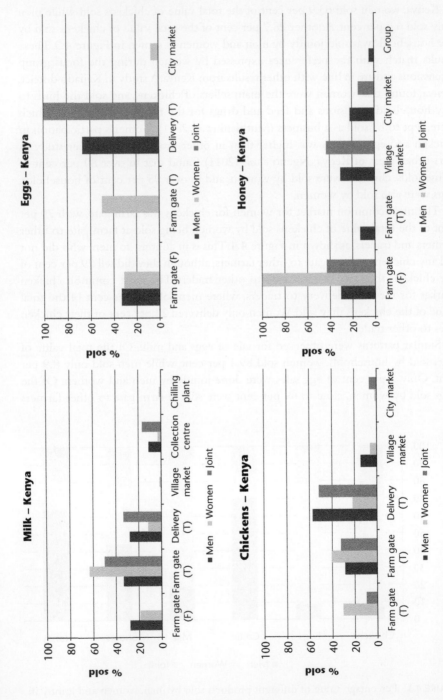

FIGURE 4.4 Types of markets where men and women sold chickens, eggs, honey and milk in Kenya

and traders, with less than 10 per cent being delivered to traders. Joint sales were all delivered to traders.

Similarly for milk: women mainly sold at farm gate to other farmers and traders. Overall, 82.3 per cent of the milk sold by women was sold through these two channels. Men, on the other hand, only sold 61.1 per cent of the milk they sold at farm gate to other farmers and traders. Women rarely delivered milk to collection centres or to traders, with less than 15 per cent of the milk they sold going to these channels. Men delivered 27.8 per cent of their milk to traders and 11.1 per cent to collection centres. Joint sales were made mainly to traders at farm gate (50 per cent), and delivered to traders (33.3 per cent) and to collection centres (16.7 per cent). Other studies have found predominant sales of milk by women. Among the Fulani of Nigeria, Waters-Bayer (1985) found that women were responsible for all milk processing and sales, including sales to village markets and door-to-door consumers. In Senegal, Dieye et al. (2005) found that milk production was entirely controlled by women, who had sole control over the sale of any surplus.

The responses to the question of who sells livestock species and products demonstrate a variation in the types of markets commonly accessed by men and women. Women were found to sell more at farm gate to other farmers and traders than to other channels that required delivery outside their homes, such as collection centres, traders and village markets. This could be due to time constraints on women, and the transaction costs involved in selling to outside markets, including costs of transport to these markets. In many cases, women do not own or control these means of transport. There is evidence that the level of women's participation diminishes as vertical integration of markets occurs, and as markets move away from sites of production and the value chain becomes more complex with multiple actors (Njuki et al. 2011b; Pionetti et al. 2011).

In Kenya, the formal dairy sector has been male-dominated due to over-reliance on cooperatives, which has limited women's participation. Marketing cooperative membership was constituted by men almost entirely because the cooperatives' function was to market produce and men controlled most cash commodities (Jacobs 1983), which include milk. In addition, some cooperatives require that members have bank accounts, through which members are paid. These dairy cooperative terms and conditions are more favourable for men than women in Kenya (Morton and Miheso 2000). Rural women are less likely than rural men to have bank accounts, making them averse to the formal dairy industry. Women may, therefore, prefer products with less formalized markets, such as goat milk, indigenous eggs, honey and manure, which are both beneficial for use at home as well as for sale in the informal markets. This tendency of women being relegated to informal markets and farm gate sales has also been noted in Tanzania (Eskola 2005). Women sometimes prefer informal markets because most rural women conduct small businesses in the informal markets in order to provide for their families, irrespective of whether they come from male- or female-headed households (Aspaas 1998). These markets provide women with more regular payments, either on a daily or weekly basis, compared to other formal channels, which have more formal and irregular modes of payment.

Joint sales were common across most of the products. This issue of "jointness" requires further research to explore what this really means.

Marketing patterns in Tanzania

In Tanzania, eggs and milk were mainly sold by women, with women selling 66.7 per cent of the total value of eggs sold and 53.3 per cent of the total value of milk sold as shown in Figure 4.5.

It is only in these two products that women sold more than men. Men sold 50.4 per cent of the total value of live chickens sold by the households, 83.3 per cent of the value of cattle sold and 75 per cent of the honey. There were more equitable sales for some commodities compared to others. For example for milk, men sold 40 per cent of the value, women sold 53.3 per cent of the value and 6.7 per cent was sold jointly. In contrast, men sold 83.3 per cent of the value of sheep and goats sold while women only sold 8.8 per cent and 7.4 per cent was sold jointly.

Farm gate to either farmers or to traders was the predominant market for sales of all commodities by men and women as well as joint sales. Similar to Kenya, women sold most of the chickens they sold at farm gate to other farmers and to traders (88 per cent) as shown in Figure 4.6. Men also sold most of the chickens they sold at farm gate (79 per cent). The key difference is who they sold to. While women sold 66 per cent to other farmers, men only sold 32 per cent of the chickens they sold to other farmers, with the rest of the farm gate sales being sold to traders. Twenty per cent of the sales by men were delivered to traders, village markets or city markets, while women only sold 12 per cent to these off-farm markets.

Most of the egg sales by men and women were made at farm gate to other farmers. Of the total value of eggs sold by women, 79 per cent was sold at farm

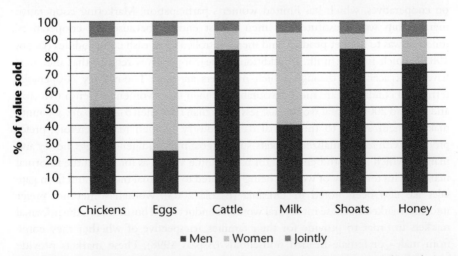

FIGURE 4.5 Percentage value of different products sold by men, women and jointly in Tanzania

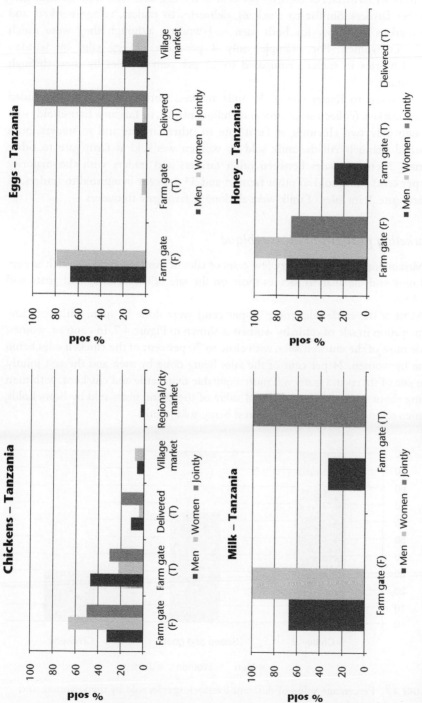

FIGURE 4.6 Types of markets where men and women sold chickens, eggs, honey and milk in Tanzania

gate to other farmers. For men, 67 per cent of the egg sales were done at farm gate to other farmers. Similar to chickens, deliveries to traders, village markets and city markets were low for both men and women, although they were much lower for women. For example, only 4 per cent of egg sales by women were deliveries to traders, compared to 11 per cent of sales by men through this channel.

In contrast to Kenya where the milk markets were diversified and included formal markets (collection centres and chilling plants), in Tanzania households sold through only two channels, at farm gate to other farmers and to traders, both informal channels. All the milk sold by women was sold at farm gate to other farmers. Men split sales between other farmers and traders with the majority (67 per cent) being sold to other farmers and 33 per cent being sold to traders, all at farm gate. Joint sales of milk were all done at farm gate to traders.

Marketing patterns in Mozambique

In Mozambique, we found very few cases of sales of livestock products such as eggs and milk and the analysis focuses more on the sale of cattle, sheep and goats, and chickens.

Most of the cattle sales (over 75 per cent) were done by men, with very low participation in sale of cattle by women as shown in Figure 4.7. In contrast, women made most of the chicken sales, with close to 70 per cent of the chicken sales being done by women, 24 per cent of the sales being done by men and the rest jointly. The sale of sheep and goats was more equitable than cattle and chickens, with men selling about 43 per cent of the total value of sheep and goats sold by households, women selling 47 per cent and the rest being sold jointly.

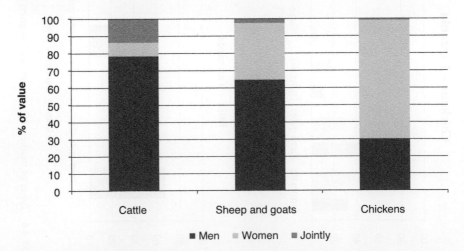

FIGURE 4.7 Percentage value of different livestock species sold by men, women and jointly in Mozambique

Women's participation in markets beyond the farm

Women were found to participate in other points of livestock value chains beyond the farm. Participatory value chain mapping showed women involved as traders of livestock products in markets, in formal and informal outlets as well as service providers.

In Kenya, two scenarios were presented by two groups of farmers on what happens to milk once it leaves the farm and women's involvement. As shown in Figure 4.8, at farm gate, milk is used for home consumption, sold to farmers or sold to traders or brokers. According to the group members, about 60 per cent of the sales at farm gate are done by women. All the brokers buying from the farmers were men. Once they bought the milk from the farmers, it went through three different channels: direct to consumers, to restaurants and shops, and to a processing plan. Farmers estimated about 50 per cent of the restaurant and shop owners were women.

The goat value chain was much shorter with more varied participation of women. Goat milk was consumed at home, sold to neighbours or sold to a collection centre. Milk sales at home to neighbours were mainly done by women (90 per cent) while sales to the collection centre were mainly done by men. Goats were sold mainly to other farmers, to brokers or retained as breeding stock. Sales to others were done by either men or women. However, the farmers felt that for the same type of goats, women got much lower prices than men did.

Conclusion

Results in this chapter show that preferences of livestock species and products, and the production and marketing of these commodities were gendered in the three countries. Men and women preferred producing commodities that they were able to market and thus control income accrued from their sales. Women were more active in the marketing of livestock products such as eggs and milk and in the marketing of small stock, especially chickens (in all three countries), and sheep and goats in Mozambique. They also participated more in farm gate markets, either to other farmers or to traders, compared to men who were more likely to sell to outside markets.

Women tend to face more challenges than men in accessing and benefiting from markets, especially more formal markets. These could include limited mobility; time poverty; lack of access to assets that would facilitate their participation, such as transport and communication assets, and bank accounts; and lack of access to market information. These constraints limit women to participating more at farm gate markets rather than markets outside their homes.

Paying attention to women's constraints to marketing by providing skills and training, increasing access to assets and technologies, and applying appropriate legal and institutional mechanisms can enable women to effectively participate in these formal value chains. In some cases, however, women are able to participate in

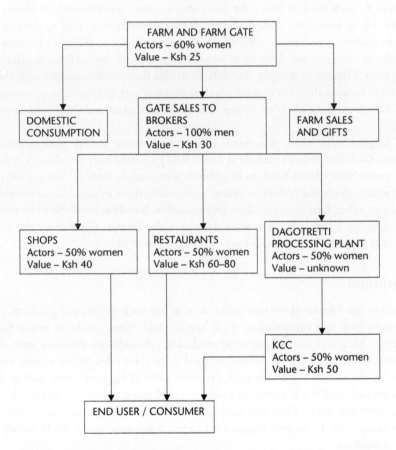

VALUE CHAIN ANALYSIS – KIAMBU
Karai United and Karai Bee Keepers

DAIRY

FARM AND FARM GATE
Actors – 60% women
Value – Ksh 25

DOMESTIC CONSUMPTION

GATE SALES TO BROKERS
Actors – 100% men
Value – Ksh 30

FARM SALES AND GIFTS

SHOPS
Actors – 50% women
Value – Ksh 40

RESTAURANTS
Actors – 50% women
Value – Ksh 60–80

DAGOTRETTI PROCESSING PLANT
Actors – 50% women
Value – unknown

KCC
Actors – 50% women
Value – Ksh 50

END USER / CONSUMER

FIGURE 4.8 Women's participation in the dairy value chain beyond the farm in Kiambu and Kajiado in Kenya

VALUE CHAIN ANALYSIS – KIAMBU
Karai Uugi and Karai Young Mothers

DAIRY

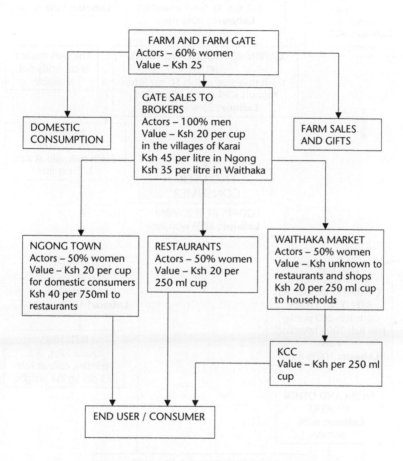

FARM AND FARM GATE
Actors – 60% women
Value – Ksh 25

GATE SALES TO BROKERS
Actors – 100% men
Value – Ksh 20 per cup
in the villages of Karai
Ksh 45 per litre in Ngong
Ksh 35 per litre in Waithaka

DOMESTIC CONSUMPTION

FARM SALES AND GIFTS

NGONG TOWN
Actors – 50% women
Value – Ksh 20 per cup
for domestic consumers
Ksh 40 per 750ml to
restaurants

RESTAURANTS
Actors – 50% women
Value – Ksh 20 per
250 ml cup

WAITHAKA MARKET
Actors – 50% women
Value – Ksh unknown to
restaurants and shops
Ksh 20 per 250 ml cup
to households

KCC
Value – Ksh per 250 ml
cup

END USER / CONSUMER

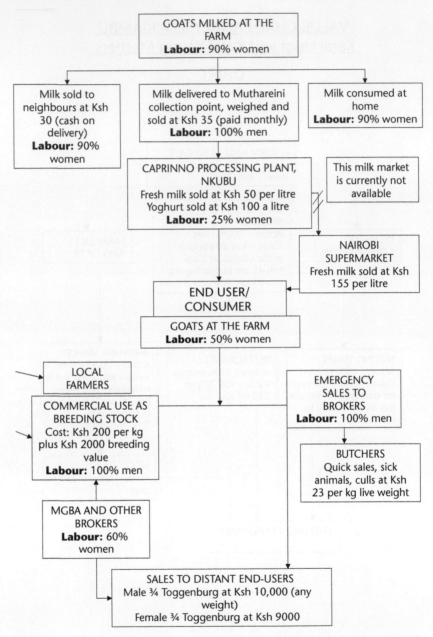

FIGURE 4.9 Women's participation in the dairy value chain beyond the farm in Meru in Kenya

different types of markets including distant regional markets. Understanding the determinants of women's participation in markets can help identify intervention areas that will optimize women's participation while optimizing their benefits. Collective action can also promote women's livelihoods and support women's empowerment. From an agriculture and markets perspective, women can pool labour, resources, assets and even marketable products to overcome gender-specific barriers that constrain them from participating in economic activities. Collective action has especially been shown to increase women's access to markets and services.

While this chapter focused mainly on women as suppliers of livestock and livestock products, some of the participatory value chain mapping showed that women were also actors in other points of the value chains as traders, restaurant and shop owners, as well as workers in collection and processing centres.

References

Aspaas, H. R. (1998) Heading households and heading businesses: rural Kenyan women in the informal sector. *Professional Geographer* 50(2): 192–204.

Brockington, D. (2001) Women's income and the livelihood strategies of dispossessed pastoralists near the Mkomazi Game Reserve, Tanzania. *Human Ecology* 29(3): 307–338.

Delgado, C. L. (2003) Rising consumption of meat and milk in developing countries has created a new food revolution. *Journal of Nutrition* 133(11): 3907S–3910S.

Duvel, G. H. and Stephanus, A. L. (2000) A comparison of economic and cultural incentives in the marketing of livestock in some districts of the northern communal areas of Namibia. *Agrekon* 39(4): 656–664.

Eskola, E. (2005) Agricultural marketing and supply chain management in Tanzania: a case study. Available at: http://www.tanzaniagateway.org/docs/agriculturalmarketingand supplychainmanagementintanzania.pdf (accessed 6 February 2011).

Fafchamps, M. and Gabre-Madhin, E. (2001) *Agricultural Markets in Benin and Malawi: Operation and Performance of Traders*. World Bank Policy Research Working Paper No. 2734 (December 2001). Washington, DC: World Bank

Fleisher, M. L. (1998) Cattle raiding and its correlates: the cultural-ecological consequences of market-oriented cattle raiding among the Kuria of Tanzania. *Human Ecology* 26(4): 547–572.

Fratkin, E. and Smith, K. (1995) Women's changing economic roles with pastoral sedentarization: varying strategies in alternate Rendille communities. *Human Ecology* 23(4): 433–454.

Gengenbach, H. (1998) "I'll bury you in the border!" Women's land struggles in post-war Facazisse (Magude District) Mozambique. *Journal of Southern African Studies* 24(1): 7–36.

Herskovits, M. J. (1924) A preliminary consideration of culture areas of Africa. *American Anthropologist* 26(1): 50–63.

Jabbar, M. A., Swallow, B. M., d'Leteren, G. D. M. and Busari, A. A. (1998) *Farmer Preferences and Market Value of Cattle Breeds of West and Central Africa.* Socioeconomic and Policy Research Working Paper No. 21. Addis Ababa, Ethiopia: Livestock Policy Analysis Program, ILRI.

Jacobs, S. (1983) Women and land resettlement in Zimbabwe. *Review of African Political Economy* 10(27–28): 33–50.

Kabeer, N. (2001) Conflicts over credit: re-evaluating the empowerment potential of loans to women in Bangladesh. *World Development* 29(1): 63–84.

Kitalyi, A. J. (1998) *Village Chicken Production Systems in Rural Africa: Household Food Security and Gender Issues*. FAO Animal Production and Health Paper 142. Rome: FAO.

Kossila, V. (1988) The availability of crop residues in developing countries in relation to livestock populations. In *Proceedings of a Workshop on "Plant Breeding and the Nutritive Value of Crop Residues" held at ILCA, Addis Ababa, Ethiopia, 7–10 December 1987*.

Kristensen, E., Larsen, C. E. S., Kyvsgaard, N. C., Madsen, J. and Henriksen, J. (2004) Livestock production: the 21st century's food revolution (Discussion paper on the donor community's role in securing a poverty-oriented commercialization of livestock production in the developing world). *Livestock Research for Rural Development* 16(1).

Kristjanson, P., Waters-Bayer, A., Johnson, N., Tipilda, A., Njuki, J., Baltenweck, I. *et al.* (2010) *Livestock and Women's Livelihoods: A Review of the Recent Evidence*. ILRI Discussion Paper No. 20. Nairobi: ILRI.

Madox, G. H. (1996) Gender and famine in central Tanzania 1916–1961. *African Studies Review* 39(1): 83–101.

McPeak, J. and Doss, C. R. (2006) Are household production decisions cooperative? Evidence on pastoral migration. *American Journal of Agricultural Economics* 88(3): 525–541.

Morrison, C. and Jutting, J. (2005) Women's discrimination in developing countries: a new data set for better policies. *World Development* 33(7): 1065–1081.

Morton, J. and Miheso, V. (2000) Perceptions of livestock service delivery among smallholder dairy producers: case studies from central Kenya. *Livestock Research for Rural Development* 12(2). Available at: http://www.cipav.org.co/lrrd/lrrd12/2/mor122.htm (accessed May 2013).

Moser, C. (2006) *Asset-Based Approaches to Poverty Reduction in a Globalized Context: An Introduction to Asset Accumulation Policy and Summary of Workshop Findings*. Washington, DC: Brookings Institution.

Muthiani, E. N., Kirwa, E. C. and Ndathi, A. J. N. (2011) Status of chicken consumption and marketing among the Maasai of Kajiado District, Kenya. *Livestock Research for Rural Development* 23(158). Available at: http://www.lrrd.org/lrrd23/7/muth23158.htm (accessed 7 July 2011).

Ngeno, V., Langat, B. K., Rop, W. and Kipsat, M. J. (2011) Gender aspect in adoption of commercial poultry production among peri-urban farmers in Kericho Municipality, Kenya. *Journal of Development and Agricultural Economics* 3(7): 286–301.

Njombe, A. P. and Msanga, Y. N. (2009) Livestock and dairy industry development in Tanzania. Department of Livestock production and Marketing Infrastructure Development, Ministry of Livestock Development. Available at: http://www.mifugo.go.tz/documents_storage/LIVESTOCK%20INDUSTRY%20DAIRY%20DEVELOPMENT%20IN%20TANZANIA%20-%20LATEST3.pdf (accessed May 2013).

Njuki, J., Waithanji, E., Kariuki, J., Mburu, S. and Lymo-Macha, J. (2011a) *Gender and Livestock: Markets, Income and Implications for Food Security in Tanzania and Kenya*. Report on a study commissioned by IDRC and Ford Foundation. Nairobi: ILRI.

Njuki, J., Kaaria, S., Chamunorwa, A. and Chiuri, W. (2011b) Linking smallholder farmers to markets, gender and intra-household dynamics: does the choice of commodity matter? *European Journal of Development Research* 23(3): 426–433.

Nori, M., Kenyanjui, M., Ahmed Yusuf, M. and Hussein Mohammed, F. (2006) Milking drylands: the marketing of camel milk in North-East Somalia. *Nomadic Peoples* 10(1): 9–28.

Nunow, A. A. (2000) *Pastoralists and Markets*. Research report 61. Leiden: African Studies.

Omore, A., Muriuki, H., Kenyanjui, M., Owango, M. and Staal, S. (1999). *The Kenya Dairy Sub-sector: A Rapid Appraisal*. Nairobi: Smallholder Dairy (Research and Development) Project.

Peeler, E. J. and Omore, A. O. (1997) *Manual of Livestock Production Systems in Kenya*, 2nd edn. Nairobi: Kenya Agricultural Research Institute.

Pionetti, C., Adenew, B. and Alemu, Z. A. (2011) Characteristics of women's collective action for enabling women's participation in agricultural markets: preliminary findings from Ethiopia. Presentation at the Gender and Market Oriented Agriculture (AgriGender 2011) Workshop, Addis Ababa, Ethiopia, 31 January–2 February 2011. Available at: http://results.waterandfood.org/handle/10568/3122 (accessed 31 December 2012).

Ridgewell, A. and Flintan, F. (2007) *Gender and Pastoralism*, vol. 2: *Livelihoods and Income Development in Ethiopia*. Addis Ababa: SOS Sahel Ethiopia. Available at: http://idl-bnc.idrc.ca/dspace/handle/10625/43750 (accessed 23 January 2012).

Rocha, A., Starkey, P. and Dionisio, A. (1991) Cattle production and utilization in smallholder farming systems in southern Mozambique. *Agricultural Systems* 37(1): 55–75.

Speedy, A. W. (2003) Global production and consumption of animal source foods. *Journal of Nutrition* 133:4048S–4053S. (Supplement: Animal Source Foods to Improve Micronutrient Nutrition in Developing Countries.)

Swetnam, J. J. (1988) Women and markets: a problem in the assessment of sexual inequality. *Ethnology* 27(4): 327–338.

USAID (2009) Women's milk and small ruminant marketing in Mandera Triangles (Kenya, Ethiopia and Somalia). Available at: http://www.elmt-relpa.org/aesito/hoapn?id_cms_doc=58...file... (accessed 31 January 2012).

van Rooyen, A. and Homann, S. (n.d.) Matabeleland's informal goat markets: their role and function in smallholder livestock development. *ICRISAT Briefing Note* No. 3.

Verhaegen, I. and Van Huylenbroeck, G. (2001) Costs and benefits for farmers participating in innovative marketing channels for quality food products. *Journal of Rural Studies* 17(4): 443–456.

Waters-Bayer, A. (1985) *Dairying by Settled Fulani Women in Central Nigeria and Some Implications for Dairy Development*. ODI Pastoral Development Network Paper 20c. London: Overseas Development Institute.

Zaal, F. (1999) *Pastoralism in a Global Age: Livestock Marketing and Pastoral Commercial Activities in Kenya and Burkina Faso*. Thela Thesis, Amsterdam.

5

LIVESTOCK MARKETS AND INTRA-HOUSEHOLD INCOME MANAGEMENT

Jemimah Njuki, Samuel Mburu and Paula Pimentel

Introduction

Allocation of resources including assets and income within households has been a focus of research since the early 1990s. Different models of resource allocation, their assumptions and limitations have been discussed by several authors including the unitary model and the collective model (Marchant 1997; Udry 1996). The unitary model has been rejected in both developed and developing countries, with important implications for policy, practice and evaluation methods (Behrman 1997; Hoddinot and Haddad 1995; Strauss and Thomas 1995).

While the influence of women-managed assets and income on development outcomes such a child nutrition, education and women's own empowerment have been studied (Quisumbing 2003; World Bank 2001), the main factors in or preconditions for women's management of income have not been studied to the same level. This chapter assumes the collective, non-cooperative, household model in which husbands and wives may pool part of their income (joint income) but retain individual incomes. Analyses of "shared" income management can clarify whether management by both women and men signifies a genuine interdependence in an environment of cooperation or male dominance in a conflicting environment. Just as in joint decision-making, joint management could represent empowerment or disempowerment due to cooperation or conflict (Kabeer 2001). In this chapter we analyse patterns of income management from sales of livestock and livestock products, and the factors that affect income control and management by women, focusing on household, intra-household and non-household factors. We especially analyse the differences in women's management of income across livestock and livestock products, the role of markets and women's participation in markets, as well as how women's income management is influenced by the amount of income going into the household. The chapter adds to the recent evidence by expanding on the models of intra-household resource allocation using data from an individual source income (in this case livestock) and through the collection of data from multiple individuals within households, in this case from men and women. As Doss (1996) points out, the estimation and use of the non–cooperative models requires much more detailed data on individual earnings and resources and how these are managed or transferred among individuals.

Intra-household decision-making and resource allocation

The unitary household model assumes that members have identical preferences and that household income is pooled (Marchant 1997; Udry 1996). Thus, individual preferences and bargaining weights for time and income allocation do not matter (Marchant 1997; Udry 1996). The cooperative bargaining model, on the other hand, perceives the household as a unit consisting of sub-units, with agency, who negotiate about the distribution of benefits, including income among households. From an income perspective, Doss (1996) distinguishes between "pooled" income, which is income that is put into a common pot and either the household head, who is assumed to be an altruist, makes decisions on the allocation, or the household bargains as to how the income should be allocated. For "non-pooled income" household members have separate incomes and individual budget constraints, and individuals may bargain over how much of their income will be joint or allocated to joint expenditure. The measurement of men and women's income in this chapter uses the concept of non-pooled income, with men and women managing distinct income, but also allocating income to some joint utility functions.

Women's income, and their ability to manage it, is especially vital to the survival of many households (Bruce 1989). Men and women play different roles in agricultural production but their participation in markets, their returns on labour and their patterns of economic participation often differ. Women's income has been associated with both individual and household benefits. Sen (1985) argues that earnings or income managed by women can provide leverage for women by offering them a fallback position in the event of a divorce, and a greater ability to deal with threats and also use threats within marriage. Women's management of income provides them with a greater bargaining power and has been shown to reduce domestic violence. The author, however, cautions that women's earning power can serve to subjugate them further, especially if men's spending on the household reduces as women manage more income,

Women's management of income has been associated with improved child nutritional status. Studies in Kenya, Botswana, Ghana, Jamaica and Guatemala (Blumberg 1988; Engle 1993; Knudsen and Yates 1981; Tripp 1981) have found a positive correlation between mothers' income and child nutritional status. Hoddinot and Haddad (1995) found an increase in anthropometric measures of children with increasing management of income my women. Similarly, Bennett (1988) found that the greater the extent of a woman's influence over the allocation of income (whether pooled or individual), the better the child's dietary intake and nutritional status was. Other evidence suggests that children benefit when women have increased access to financial assets (Green and White 1997), just as they do better when their mothers control a larger percentage of family income. When women control a larger fraction of the family financial resources, more resources are allocated for children's needs.

It therefore follows that when women lose control of income, what is affected is not only their relative marital/familial power (and self-esteem) but also family well-being (Blumberg 1988). Early studies by Hanger and Morris (1973) have explored this negative relationship and found that a reduction in women's income brought about by development projects that aim to maximize women's labour inputs into male-controlled crops had a negative impact on children's nutrition. These trends are mainly as a result of differences in expenditure patterns between men and women. Early studies on men and women's expenditure patterns of income under their control showed women spent almost 100 per cent of their income on family while men only spent a portion of their income on the family, even when the overall income was not sufficient to meet family needs. These patterns may be defined by the perceived traditional roles of men and women, where men provide such items as housing and schooling while women are responsible for the food, nutrition and health of the family.

More often than not, the management of income by women is not commensurate with the time allocated to the activities from which the income is derived. The view of men as heads of households who should be responsible for income earned in the household and women's weaker bargaining power all play a part in reducing the amount of income that women manage or control, especially from family and/or joint activities within the household such as agriculture production. Even when women have control over such income (e.g. from women's cropped plots, small livestock, own earnings), the commercialization of production can erode this control. There is some evidence that, as agricultural products become commercialized, women have often lost control and management of the income derived from these products (Katz 2000; Kennedy and Cogill 1987; Njuki *et al.* 2011; von Braun 1995; von Braun and Webb 1989; von Braun *et al.* 1989).

Despite this association of women's management of income with important development outcomes, the *World Development Report* (World Bank 2012) documents that in some countries many women continue not to make decisions even on their own income. For example as many as 34 per cent of married women in Malawi and 28 per cent of women in the Democratic Republic of Congo are not involved in decisions about spending their earnings. And 18 per cent of married women in India and 14 per cent in Nepal are largely silent on how their earned money is spent. Within agriculture, studies on women's management of income have mainly been done from crop production activities (Kennedy and Cogill 1987; Njuki *et al.* 2011; von Braun 1995). With the projected increased demand for livestock and livestock products especially for urban consumers (Delgado 2003), it is expected that the markets and marketing systems for livestock and livestock products are changing and will continue to change in the near future. How this will affect the management and control by women, especially of livestock products such as milk and eggs, has not been studied. And while studies have been done on the extent to which women manage income and the impacts of technology and commercialization on this, the factors that influence women's management of income have been less studied.

Often the interventions to increase benefits from markets, be they agricultural or non-agricultural, have shied away from trying to change intra-household dynamics that influence income management, due to a lack of clarity on how best to influence intra-household income and resource allocation. Those working on the development of or research on these interventions either assume a unitary model, where the main objective is to increase the income of the household or the head of the household, with the assumption that this will lead to the income being distributed within the family without a consideration of the degree to which this income will leak out to other purposes or be allocated for the benefit of the welfare of all household members. While there has been concern about the distributional impact of interventions, especially market-led interventions, this has often focused on class disparities rather than within-household disparities. One of the approaches that development programs have taken is to focus on targeting strategies, such as working with women's groups, anticipating that they will increase income under the management of women from such approaches.

Contribution of livestock to household income

Sale of milk and chickens provided the largest amount of income in Kenya, while sale of cattle generated the largest amount of income in Mozambique (see Figure 5.1). In Tanzania the amount of income from different livestock and livestock products did not vary greatly. In Kenya milk was an important source of income, contributing up to 40 per cent of all the livestock income and 29 per cent of all the household income. Sale of sheep and goats was the second largest contributor to both total livestock and household income. In contrast, the largest contributor to income in Mozambique was sale of cattle (37 per cent of livestock income and 31 per cent of household income), followed by sale of sheep and goats (31 per cent of livestock income and 25 per cent of household income). In Tanzania the largest contributor to livestock income was sale of chickens, which was 31 per cent of the total livestock income and 7 per cent of the total household income. Looking across the three countries, livestock as a source of income were least important in Tanzania, where none of the livestock species or products contributed more than 10 per cent of the household income.

Patterns of income management across livestock and livestock products

While women's roles in livestock production and marketing differ from one production system to another, from region to region and country to country, women do provide most of the labour in livestock in sub-Saharan Africa. The

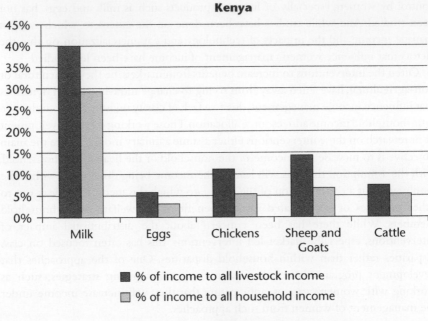

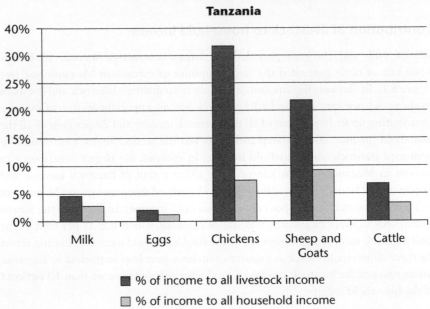

FIGURE 5.1 Contribution of livestock and livestock products to household income in Kenya, Tanzania and Mozambique *(Continued)*

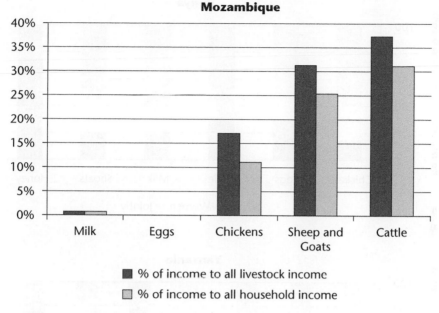

FIGURE 5.1 *(Continued)*

control of income from these livestock and their products also differs depending on the type of livestock and the type of livestock product, among other things. There is a common perception that women are more likely to own small stock, such as chickens, sheep and goats, rather than larger animals, such as cattle, water buffaloes and camels, and therefore will benefit more from small stock than from the larger stock (Kristjanson *et al.* 2010). Studies have shown, however, that women may manage income from sale of livestock products even when they do not own the livestock itself. For example Waters-Bayer (1988), in a study in Nigeria, found that although women did not own cattle, they controlled and managed income from the sale of milk.

In Tanzania, women managed more income from the sale of small livestock than large livestock. For example, they managed 49 per cent of income from the sale of chickens and 33 per cent of income from the sale of sheep and goats compared to 24 per cent of income from the sale of cattle. On management of livestock and their products, women managed 50 per cent of the income from the sale of milk, which was much higher than their income share from the sale of cattle (24 per cent). This pattern was not true for chickens and eggs, however. While women managed 49 per cent of the income from chickens, they only managed 29 per cent of the income from eggs. Most of the income from eggs (67 per cent) was managed jointly. There are several reasons for the patterns on management of eggs, the main one being the commercial nature of the egg markets in both Kenya and Uganda, where there is a growing demand in the larger cities and therefore eggs are produced more for the urban population than for local rural consumers. This commercialization has led to the egg business becoming more and more male dominated.

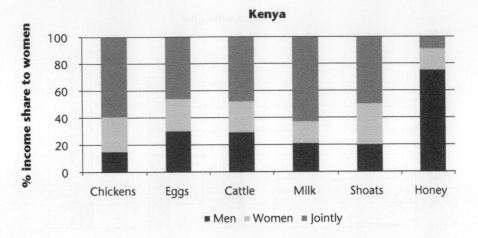

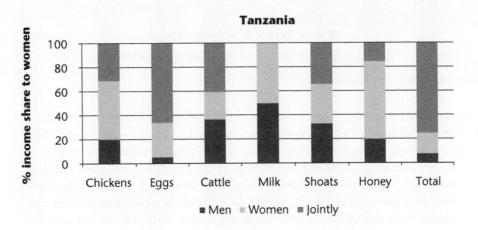

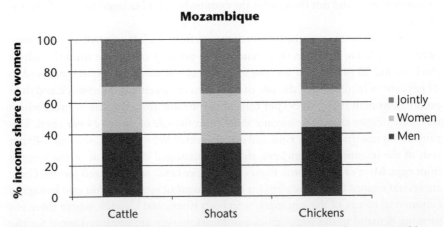

FIGURE 5.2 Proportion of livestock and livestock product income share managed by women in Tanzania, Kenya and Mozambique

In Kenya, most of the income was reported as jointly managed. There was no significant difference in the proportion of income managed by women from the large and small stock. They managed almost the same proportion of income from the sale of chickens, cattle, and sheep and goats (26 per cent, 22 per cent and 30 per cent respectively). Analysing across different livestock species and their products, women managed 26 per cent of income from chickens compared to 24 per cent of income from the sale of eggs. In Mozambique, there was almost equal management of income between men and women, and joint management for cattle and goat sales. In contrast to the other two countries, women managed a smaller proportion of the income from chickens than from cattle, sheep and goats. From focus group discussions in Mozambique it was evident that women are very involved in the marketing of livestock, especially goats, whereby there exists a thriving trade in goats, mainly controlled by women traders who purchase goats from the district and bring them into the capital city, Maputo, by train.

Influence of types of markets on women's management of income

The type of market that a product is sold to has been shown to influence the income share going to women (Njuki et al. 2011). There is evidence that women are more likely to sell to informal, often near-to-home markets, and that income derived from these markets will be managed by women. There were several markets that livestock and livestock products were sold to: farm gate to other farmers, farm gate to traders, village markets, and delivery to shops/traders/butchers and other market actors.

Generally, women are expected to manage a larger income share when products are sold in informal markets, often at farm gate, compared to when they are sold in distant markets, or when delivered under contract or other arrangements to formal establishments such as shops and butcheries. At farm gate level, it was expected that women would sell more and therefore manage more income if products were sold to other farmers than when sold to traders, often due to their lower negotiation skills and social capital from interactions with other farmers.

In Tanzania, when chickens were delivered to traders and shops away from home, women lost up to 35 per cent of the income share that they would have managed if they sold chickens at farm gate to other farmers (Figure 5.3). When chickens were sold at farm gate to other farmers, women received a 70 per cent share of the income. This share, however, fell to 45 per cent when chickens were sold at farm gate to other traders, and further to 28 per cent when the chickens were delivered to traders, shops or hotels. Similar trends were observed for other products. For milk, when sold at farm gate to other farmers, women's income share was 74 per cent. This fell by more than 50 per cent to 32 per cent when the milk was delivered to traders, shops or hotels. At farm gate, selling eggs to traders instead of to other farmers, reduced women's income share by 24 per cent. This trend was not observed, however, for the sale of cattle, sheep and goats, which did not seem to be greatly influenced by the type of market.

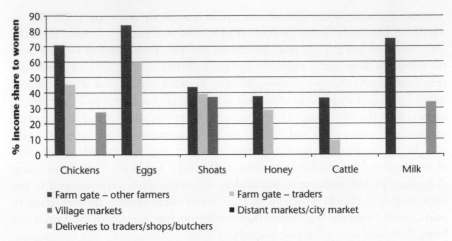

FIGURE 5.3 Percentage income share to women based on where livestock was sold in Tanzania

Similar to Tanzania, in Kenya women managed the highest proportion of income from chickens and eggs when these products were sold at farm gate to other farmers (Figure 5.4). Selling eggs at farm gate to traders and not other farmers also reduced the income share going to women by close to 20 per cent. Unlike in Tanzania, however, the proportion of chicken income managed by women was much higher when chickens were sold to village markets than when sold at farm gate to other traders.

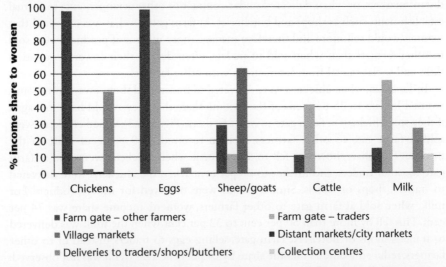

FIGURE 5.4 Percentage income share to women based on where livestock was sold in Kenya

Income management patterns were less clear for the sale of cattle, sheep, goats and milk. Women managed the largest income share from the sale of sheep and goats when these were sold in the village market, and from the sale of cattle if cattle were sold at farm gate to other traders. This could be due to lower sales of these species among farmers. In Mozambique the most common sales were of chickens, cattle, sheep and goats. There was no reported sale of milk and eggs, either at farm gate or through other channels. Similar to Kenya and Tanzania, women managed the highest proportion of income from sales at farm gate to other farmers for chickens, sheep and goats (see Figure 5.5).

Influence of women's participation in markets on their income management

Women's participation in market transactions can influence the extent to which they manage income. Often, development programs focus on increasing access to markets by women to enhance their benefits and management of income from these market linkages. In Tanzania, this was true across all the species and products. For chickens and milk, women managed close to 100 per cent of the income when they sold the chickens and milk themselves, compared to only 26 per cent and 17 per cent for chickens and milk respectively when these products were sold by men. Even in the case of sheep and goats, women managed 60 per cent of the income when they made the actual sale, compared to 35 per cent when men made the actual sale (Figure 5.6).

These results show the value of linking women farmers to markets so that they are able to do the negotiations and carry out the transactions themselves. Women

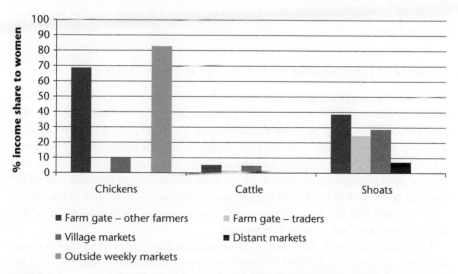

FIGURE 5.5 Percentage income share to women based on where livestock was sold in Mozambique

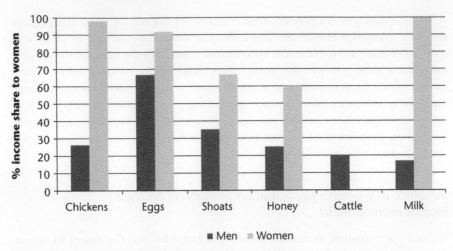

FIGURE 5.6 Percentage income share to women depending on who sold livestock in Tanzania

often face different constraints in participating in markets, including issues of mobility, balancing household reproductive and care work with market participation, access to information and infrastructural facilities in markets, low literacy and negotiation skills. While development programs aiming to increase benefits to women through markets have focused on addressing these constraints, these results show that if addressing these constraints can facilitate women's direct participation in markets they will also influence intra-household income management and resource allocation in favour of women. Such programs, however, need to work with men as what little income women control may substitute for former male household member contributions if men retain more of their income for their own individual use.

Influence of total incomes on the income share to women

There is both anecdotal and documented evidence on changes in control of products or enterprises once they become more commercialized or successful, or once the total income from these products becomes large (Njuki *et al.* 2011). Although the study did not collect time series data on change in income management over time, we use different species and correlate the total amounts received from each and the income share that goes to women. In Kenya, milk which was the livestock enterprise that paid the most had the lowest share of income managed by women, while sheep, goats and eggs had the lowest amounts of income and a higher proportion of the income from these was managed by women.

In Tanzania, although there was no linear relationship between the total amount of money made from different livestock and livestock products and the income share managed by women, products or livestock that had high incomes (with the exception of milk) also had the lowest share of income managed by women

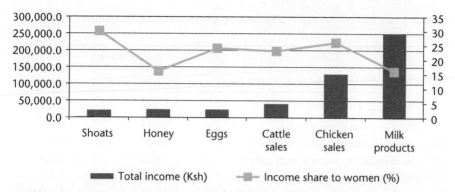

FIGURE 5.7 Relationship between total income and income share managed by women in Kenya

(products on the left side of Figure 5.8 such as eggs and cattle). Livestock and livestock products with lower amounts of total income (i.e. sheep and goats, honey and chickens), had higher income shares managed by women.

Mozambique had too few livestock and livestock products sold to enable a comparison of total income and income share managed by women across species and products.

Influence of household, intra-household and non-household factors in determining women's management of livestock income

Other intra-household, household and community factors can influence women's management of livestock income. Ownership of assets (including human capital, land, domestic assets and livestock) has been shown by various studies to increase

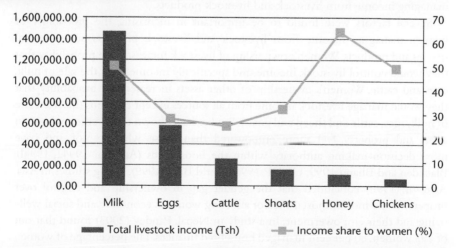

FIGURE 5.8 Relationship between total income and income share managed by women in Tanzania

TABLE 5.1 Factors influencing management of livestock income by women for all species and products

Women manage livestock income (1 = yes, 0 = no)	All livestock income	
	Coefficient	t-value
Size of land owned	0.021	1.96*
Has woman received training (yes, no)	0.033	0.17
Does woman belong to a group (yes, no)	0.076	0.39
Other assets owned by women (asset index)	0.012	−1.86*
Education level	−0.101	−1.3
Women's ownership of livestock (proportion of Tropical Livestock Units [TLUs] owned by women)	1.513	3.62***
Whether delivered to buyers (1 = yes)	−0.300	−1.15
Whether sold to village markets (1 = yes)	−0.344	−1.66*
Total household income (except livestock)	0.000	−1.2
Constant	−0.075	−0.3
Number of observations	244	
Design df	243	
F(10, 234)	2.55	
Prob. > F	0.0062	

***, **, * significant at 1%, 5% and 10% respectively.

women's decision-making and women's empowerment (Kabeer 2001). Ownership of livestock by women may give them more decision-making authority on the disposal or the use of the livestock products as well as management of the income. We ran a probit analysis to analyse the factors that increase the probability of women managing income from livestock and livestock products.

Several factors were found to be important in increasing the probability of women managing income from livestock and livestock products across several species and products. Women's ownership of livestock increased the probability that they would control livestock income, and specifically income from the sale of milk, eggs and cattle. Women's ownership of other assets increased the probability that they would manage livestock income from all sources except chickens (i.e. from sale of milk, eggs, cattle and sheep and goats). Studies have shown that women who own assets and property feel more empowered than those who do not and have more decision-making authority within the households (Agarwal 1994a, 1994b; Blackden and Bhanu 1999; UNDP 1996; World Bank 1999). Using studies in Asia, Agarwal (1998) concluded that the gender gap in ownership and control over property is the most important factor affecting women's economic and social well-being and their empowerment. In a study in Nepal, Pandey (2003) found that out of 293 women, 80 per cent managed household finances. The percentage of women who managed finances was, however, much higher for women in households where men and women owned property (92 per cent), and when women were sole owners (86 per cent) compared to households where husbands were sole owners

TABLE 5.2 Factors influencing women's management of income from livestock products (milk and eggs)

Women manage income (1 = yes, 0 = no)	Women's management of milk income		Women's management of income from eggs	
	Coefficient	t	Coefficient	t
Land owned	0.009	0.22	0.033	1.13
Received training (1 = yes)	0.436	1.17	−0.059	−0.1
Household size	0.140	2.44★★	−0.855	−1.43
Milk traders (1 = yes)	0.003	1.54	0.001	1.16
Milk delivered to buyer (1 = yes)	0.000	−1.59	0.002	−3.07★★★
Proportion of spouse TLUs to total TLUs	3.396	3.39★★★	5.698	2.88★★★
Other incomes (except livestock)	0.000	0.85	0.001	2.79★★★
Other assets owned by women	0.043	−2.34★★	−0.034	−2.91★★★
Kenya (1 = yes)	−0.574	−1		
Constant	−0.438	−0.62	−0.234	−0.25
Number of observations	110		48	
Design df	109		47	
F(10, 100)	2.6		2.95	
Prob. > F	0.0075		0.0077	

★★★, ★★, ★ significant at 1%, 5% and 10% respectively.

(79 per cent). Increasing women's control over land, physical assets and financial assets can improve child health and nutrition, and increase expenditures on education, contributing to overall poverty reduction (Meinzen-Dick et al. 2011).

The location where the sale was made influenced the probability of women managing income. As discussed earlier, women were more likely to manage livestock income when sales were made at farm gate compared to when products or livestock were delivered to traders outside the farm or sold at village markets. Women were less likely to manage income from eggs, chickens and cattle if these were delivered to traders compared to if they were sold at farm gate or in village markets. Results from Kenya, Uganda and Tanzania have shown similar trends in the marketing of crops (Njuki et al. 2011). Results on the role of women making market transactions or doing sales direct influencing their ability to manage income have been discussed above. There are several reasons for women's higher participation in farm gate sales and their income management from these types of sales compared to those made away from home. Where women are unable to transport livestock and livestock products to market, men generally make the financial transactions and retain the income. Their roles therefore tend to diminish as the formal markets expand unless strategies are pursued that ensure they participate in these markets. The need to balance their reproductive roles and market-based roles may, however, limit their participation, with most of their sales being made at farm gate where market participation can be combined with other household activities. Although in many cases these farm gate markets may not be as profitable, they are important in diversifying income sources for women, building their confidence in dealing with markets and providing them with much needed cash flow that they can manage and control.

In wealthier households, women were expected to be managing more income from livestock than in poorer households. Research has shown that where there are multiple sources of income, women are more likely to manage income from some

TABLE 5.3 Factors influencing women's management of income from livestock sales (chickens, cattle, sheep and goats)

Women manage income (1 = yes, 0 = no)	Women's management of income from chicken sales		Women's management of income from cattle sales		Women's management of income from shoat sales	
	Coeff.	t	Coeff.	t	Coeff.	t
Land owned	0.004	0.47	−0.005	−0.42	−0.018	−1.58
Belong to group (1=yes)	0.493	2.51**	0.157	0.42	0.018	0.08
Household size	0.064	1.94*	0.017	0.37	0.007	0.19
Delivered to traders (1 = yes)	0.001	−2.03**	0.002	−1.85*	0.001	−0.59
Village market (1 = yes)	0.001	0.79	0.001	−1.54	0.001	−0.33
Proportion of TLUs belonging to women	−0.063	−0.23	0.684	1.77*	0.323	1.03
Other incomes (except livestock)	0.001	−0.88	0.002	0.81	0.001	−1.38
Other assets belonging to women	0.007	0.9	0.049	2.95***	0.108	4.08***
Kenya (1 = yes)	0.278	0.91	0.049	2.95***	−0.261	−0.8
Tanzania (1 = yes)	−0.166	−0.66	1.401	2.79***	0.571	1.73*
Constant	−0.231	−0.71	−1.356	−1.28	−0.673	−0.85
Number of observations	234		139		179	
Design df	233		138		178	
F(11, 223)	1.87		2.29		2.4	
Prob. > F	0.0446		0.0139		0.0068	

***, **, * significant at 1%, 5% and 10% respectively.

of these sources (Njuki *et al.* 2011). In wealthier households, women were more likely to manage income from the sale of livestock products (eggs) but not the livestock itself.

Women use social capital for different purposes: as a social mechanism, for accumulation of assets and for accessing markets, among others. Belonging to a group increased the probability that women would manage income from the sale of some livestock species such as chickens. Social capital has been shown to give women voice and to offer opportunities for women to save money and access credit which, in turn, empowers women. Mayoux (2002), refers to these as the virtuous spirals, that as women through social capital, asset accumulation and income control change, their role in household decision-making also increases. Women playing a greater economic role can in turn transform gender relations at the household and community level, leading to further asset and income accumulation by women. Social capital, and the benefits that come with it, can be a strong catalyst for this spiral of change. Groups can increase women's access to information and support women's economic activity.

Conclusion

The study finds that women's management of income from livestock and livestock products differs both across products and across species. Women managed more income from small livestock and livestock products than from the larger stock in some of the studied countries. The types of markets that livestock and livestock products are sold to, and who they are sold to in these markets, influences women's management of income. When livestock and livestock products are sold at farm gate to other farmers, women manage a significantly higher proportion of income compared to when the livestock and products are sold in village markets or delivered to traders. For some of the livestock and livestock products, selling in marketing channels other than at farm gate reduced income share going to women by over 20 per cent. Not all farm gate sales, however, lead to management of income by women. Due to women's low negotiation skills, sales to traders at farm gate are often done by men and most of the income from such sales goes to them, which means lower income shares for women. This is not to say that women should be confined to farm gate sales to other farmers, which in a lot of cases means lower prices as the scope for negotiation and access to information from the market are lower, but any interventions to improve marketing of livestock and livestock products, such as formalizing milk marketing through cooperatives, should take account of the lower income share that women manage from these markets and put in place appropriate mechanisms to ensure that women do not lose their management of income with such market changes.

The participation of women in marketing influences the proportion of income that they manage. Various development programs seeking to link smallholder farmers to markets have focused on how to get women to participate more in market-oriented agriculture as well as increase the benefits to markets. This in turn will influence the intra-household income management which is more often difficult to influence directly. The strategy of linking women directly to markets, while it has the potential to lead to these intra-household income allocation changes in favour of women, should, however, take into account the balance between women's reproductive and care work and the market work as well as the danger of men misallocating funds to their own individual uses once women start managing higher income share from agricultural marketing. These programs would need to work with both men and women to ensure that both participate and benefit from these programs.

Market participation by women and their income management is also influenced by their ownership of the livestock itself. Although there is evidence of women marketing and managing income from sale of products such as milk, even when they do not own the cattle, the ownership of cattle by women increases significantly the likelihood that women will manage income from their sale. Ownership of assets such a livestock does not, however, only increase the likelihood that women will manage income, but also has positive consequences for other development outcomes such as nutrition and the education of children, due to the different expenditure patterns between men and women and due to the empowering nature of asset ownership. Programs targeting increasing women's ownership of livestock have potential to reduce the gender asset gap found in many developing countries.

Given the assumption of the collective households where there is choice whether or not to pool income and where the more likely scenario is both individual and pooled income, data on market participation, income earnings and income management need to be collected at individual level. This needs to go beyond collecting individual-level data that is reported by one member of the household to interviewing both men and women, especially in households with both a male and a female adult. It is only in this way that the full extent of intra-household income and resource distribution issues can be better understood and strategies designed to address them.

References

Agarwal, B. 1994a. *A Field of One's Own: Gender and Land Rights in South Asia*. New York: Cambridge University Press.

Agarwal, B. 1994b. Gender and command over property: a critical gap in economic analysis and policy in south Asia. *World Development* 22(10): 1455–1478.

Agarwal, B. 1998. Widows versus daughters or widows as daughters? Property, land, and economic security in rural India. *Modern Asian Studies* 32(1): 1–48.

Behrman, J. 1997. Intrahousehold distribution and the family. *Handbook of Population and Family Economics* 1: 125–187.

Bennett, L. 1988. The role of women in income production and intra-household allocation of resources as a determinant of child nutrition and health. *Food and Nutrition Bulletin* 10(3): 16–26.

Blackden, C. M. and Bhanu, C. 1999. *Gender, Growth, and Poverty Reduction*. Special Program of Assistance for Africa 1998, Status Report on Poverty, World Bank Technical Paper No. 428. Washington, DC: World Bank.

Blumberg, R. 1988. Income under female versus male control: hypotheses from a theory of gender stratification and data from the Third World. *Journal of Family Issues* 9: 51–84.

Bruce, J. 1989. Homes divided. *World Development* 17(7): 979–991.

Delgado, C. 2003. Rising consumption of meat and milk in developing countries has created a new food revolution. *Journal of Nutrition* 133: 3907–3910.

Doss, C. 1996. Testing among models of intra-household resource allocation. *World Development* 24: 1597–1609.

Engle, P. L. 1993. Influences of mothers' and fathers' income on children's nutritional status in Guatemala. *Social Science & Medicine* 37(11): 1303–1312.

Green, R. and White, M. 1997. Measuring the benefits of homeowning: effects on children. *Journal of Urban Economics* 41(3): 441–461.

Hanger, J. and Morris, J. 1973. Women and the household economy. In R. Chambers and J. Morris (eds) *Mwea: An Irrigated Rice Settlement in Kenya*. Munich: Weltforum Verlag, pp. 209–244.

Hoddinott, J. and Haddad, L. 1995. Does female income share influence household expenditure? Evidence from Ivory Coast. *Oxford Bulletin of Economics and Statistics* 57(1): 77–96.

Kabeer, N. 2001. Conflicts over credit: re-evaluating the empowerment potential of loans to women in rural Bangladesh. *World Development* 29: 63–84.

Katz, E. 2000. Does gender matter for the nutritional consequences of agricultural commercialization? Intrahousehold transfers, food acquisition, and export cropping in Guatemala. In A. Spring (ed.) *Women Farmers and Commercial Ventures: Increasing Food Security in Developing Countries*. Boulder, CO: Lynne Rienner, pp. 209–232.

Kennedy, E. and Cogill, B. 1987. *Income and Nutritional Effects of the Commercialization of Agriculture in Southwestern Kenya*. IFPRI Research Report No. 63. Washington, DC: IFPRI.

Knudson, B. and Yates, B. 1981. *The Economic Role of Women in Small-scale Agriculture in the Eastern Caribbean: St. Lucia*. Barbados: Women in Development Unit, University of West Indies.

Kristjanson, P., Waters-Bayer, A., Johnson, N., Tipilda, A., Njuki, J., Baltenweck, I. *et al.* 2010. *Livestock and Women's Livelihoods: A Review of the Recent Evidence*. Discussion Paper No. 20. Nairobi: ILRI.

Marchant, M. (1997) Bargaining models for farm household decision making. *American Journal of Agricultural Economics* 79: 602–604.

Mayoux, L. 2002. Microfinance and women's empowerment: rethinking best practice. *Development Bulletin* 57: 76–81.

Meinzen-Dick, R., Behrman, J., Menon, P. and Quisumbing, A. 2011. *Gender: A Key Dimension Linking Agricultural Programs to Improved Nutrition and Health*. 2020 Conference Brief 9. Washington, DC: IFPRI.

Njuki, J., Kaaria, S., Chamunorwa, A. and Chiuri, W. 2011. Linking smallholder farmers to markets, gender and intra-household dynamics: does the choice of commodity matter? *European Journal of Development Research* 23: 426–443.

Pandey, S. 2003. *Assets Effects on Women: A Study of Urban Households in Nepal*. Working Paper No. 03-04. Washington, DC: Centre for Social Development, George Washington University.

Quisumbing, A. R. (ed.) 2003. *Household Decisions, Gender, and Development: A Synthesis of Recent Research*. Washington, DC: IFPRI.

Sen, A. 1985. Well-being, agency and freedom: the Dewey lectures 1984. *Journal of Philosophy* 82: 169–221.

Strauss, J. and Thomas, D. 1995. Human resources: empirical modelling of household and family decisions. *Handbook of Development Economics* 3: 1883–2023.

Tripp, R. 1981. Farmers and traders: some economic determinants of nutritional status in northern Ghana. *Journal of Tropical Pediatrics* 27(1): 15–22.

Udry, C. 1996. Gender, agricultural production, and the theory of the household. *Journal of Political Economy* 104: 1010–1046.

UNDP 1996. *Human Development Report 1996.* New York: Oxford University Press.

von Braun, J. 1995. Agricultural commercialization: impacts on income and nutrition and implications for policy. *Food Policy* 20(3): 187–202.

von Braun, J. and Webb, P. 1989. The impact of new crop technology on the agricultural division of labor in a West African setting. *Economic Development and Cultural Change* 37(3): 513–534.

von Braun, J., Puetz, D. and Webb, P. 1989. Irrigation technology and commercialization of rice in the Gambia: effects on income and nutrition. IFPRI Research Report No. 75. Washington, DC: IFPRI.

Waters-Bayer, A. 1988. *Dairying by Settled Fulani Agropastoralists: The Role of Women and Implications for Dairy Development.* Kiel: Vauk Wissenschaftsverlag.

World Bank 1999. Mainstreaming gender and development in the World Bank: progress and recommendations. Available at: http://www.worldbank.org/publications/maingend/maingend.html (accessed 29 August 1999).

World Bank 2001. *Engendering Development: Through Gender Equality in Rights, Resources, and Voice.* Washington, DC: World Bank.

World Bank 2012. *The World Development Report 2012: Gender Equality and Development.* Washington, DC: World Bank.

6

WOMEN'S ACCESS TO LIVESTOCK INFORMATION AND FINANCIAL SERVICES

Samuel Mburu, Jemimah Njuki and Juliet Kariuki

Introduction

Information is an economic resource, and "information poverty" is increasingly being recognized as one of the prime causes of underdevelopment (Chowdhury 2006; Romer 1993). Access to information is more likely to be limited for those who are already marginalized – by their limited access to other resources, by their location in remote rural areas, or by their gender. The United Nations (UN) considers that after poverty and violence, the third major challenge facing women in developing countries is lack of access to information (Primo 2003).

Why is access to information so important? Households whose access to information is either limited, or very costly, may be unaware of other resources available to them, may fail to allocate their resources efficiently, may forgo income-enhancing opportunities, or may bear unnecessarily high levels of risk. This would be the case if, for example, individuals are unaware of several forms of information, including the requirements for obtaining loans with favourable conditions, how to obtain land titles, existing markets for their products, available technologies that could increase their profits, and how to insure themselves against idiosyncratic shocks. In the specific context of financial markets, inadequate access to information can lead producers to choose a suboptimal loan, savings or insurance strategy despite the options available to them, or to simply abstain from participating in formal financial markets (Stango and Zinman 2008).

Addressing the challenges faced by the livestock sector depends increasingly on an effective and efficient flow of information. This is crucial to addressing the production, economic, environmental and health aspects, among others, of the sector. Whether on a small or a large scale, women and men producers and processors depend on information related to markets, consumer demands and disease patterns to help them plan their enterprises. The lack of transparent, timely and reliable livestock marketing information is seen by many as one of the greatest challenges to the development of the livestock industry in East Africa. Without access to information about livestock prices at different regional markets, livestock keepers cannot identify which points-of-sale offer the best prices for their livestock and livestock products.

While information is increasingly recognized as an important resource for development, there is little empirical evidence on the extent of information poverty in the rural areas of developing countries (Chowdury 2006). In particular, there is scanty sex-disaggregated evidence documenting how women's access to livestock information and financial services compares to men's. This chapter analyses the intra-household disparities in access to livestock information and financial services among rural households using data from selected districts in Kenya. Specifically, the analysis compares women's access to information on livestock production and financial services with that of men. Women's information access matters for several reasons. If women's access to information is more limited or more costly than that of men of similar backgrounds, women may either have less access to economic opportunities or have limited engagement in the optimal use of the resources they control. Rural financial services help households to increase their incomes and build the assets that allow them to mitigate risks, smooth consumption, plan for the future, increase food consumption, and invest in education and other welfare-related needs. The data presented in this chapter identifies systematic differences between women and men's access to information on livestock production and marketing, knowledge and access to financial services. The data also looks at women's accumulation of savings and the factors that influence this.

Market information services and their role in influencing market participation

There is increased recognition of the role of market information in making marketing more efficient and equitable. Market information systems have been developed to cater for different sectors including the livestock sector. Market information systems are information systems used in gathering, analysing and disseminating information about prices and other information relevant to farmers, livestock keepers, traders, processors and others involved in handling agricultural products. Market information systems play an important role in agro-industrialization and food supply chains. These systems could take many forms, traditional or ICT (information and communication technologies) based. With the advance of ICTs for development in developing countries, the income-generation opportunities offered by market information systems have been sought by international development organizations, non-governmental organizations (NGOs) and businesses alike.

In Kenya, the National Farmers Information Services (NAFIS) is a comprehensive information service intended to serve farmers' needs throughout the country, including the rural areas where internet access is limited. It enables farmers get critical extension information through the internet or phone. The Livestock

Information Network and Knowledge System (LINKS) is a livestock marketing information system designed to provide marketing information, particularly in the major livestock-producing areas in the arid and semi-arid lands of Kenya. LINKS was designed to respond to livestock marketing information needs by providing an ICT infrastructure for reporting and requesting information on livestock prices and volumes from a network of different markets. Other initiatives to provide agriculture and livestock information include the International Development Research Centre (IDRC) funded community-based telecentre program, the ACACIA initiative, which was designed as an integrated program of demonstration projects and research to advance the access of disadvantaged communities in Africa to modern ICTs and to apply them to their own development priorities (IDRC 1998). The main aim was to provide major improvements to rural communities' access to information and ease of communications.

Access to information can be critical to increasing benefits for farmers from livestock markets. Market information systems help to attain efficient or competitive markets through reduction of information asymmetries among food system participants, which leads to reduction in transaction costs. Access to information by market actors also helps to level the playing field for all the actors, especially those who cannot meet the costs of accessing information (Azzam and Schroeter 1995).

Kizito (2011), in an analysis of agricultural market information systems, found that reception of improved agricultural market information was influenced by farmers' involvement in the production of marketable staples, access to alternative ICTs, and access to markets and extension services. The author found that holding all factors constant, reception of market information increased farmers' probability of market participation by 34 per cent. Farmers with access to information are more likely to get higher prices than those without information. In Mozambique, Kizito (2011) found the average price difference per kilogram of maize sold between households with and without information (also referred to as an information premium or information rent) was 12 per cent.

Rural women's access to information

Women and men have different access to markets, infrastructures and related services. For the most part, women producers face greater constraints than men in accessing different points along livestock value chains, as well as the related technologies, infrastructures and information about livestock markets. A study undertaken by the International Food Policy Research Institute (IFPRI) in Ethiopia showed that an increase of 10 km in the distance from the rural village to the closest market town reduces the likelihood of sales of livestock and livestock products, and decreases the likelihood that women will engage in and sell processed foods (Dercon and Hoddinott 2005). Women who lack financial capital also have a more difficult time accessing privatized veterinary and extension services that are often essential in helping producers meet market standards. One example of how this could happen comes from a study in Orissa, India (IFAD 2004) where, although dairy cooperatives were established in the wives' names, a committee of men actually managed the group. Along with traditional veterinary and extension services, women's networks and groups have been proven to be useful "organizational" pathways for passing information on livestock to women. A study on Heifer Project International's efforts to disseminate improved goat breeds through a village group process in Tanzania showed that social capital influenced people's ability to access a goat. Their ability to access and manage information was also crucial (De Haan 2001).

In spite of the growing recognition of information and knowledge as critical determinants of economic performance, access to timely, relevant and affordable information in the rural communities of developing countries remains very limited. The reasons have much to do with lacking, poorly developed or poorly maintained infrastructure; rural dwellers' significantly lower income levels; and the lack of information content that is targeted to local needs (Munyua 2000). For women within these rural communities, these constraints are compounded by socially constructed gender roles and relationships that further hinder women's ability, relative to men, to access information. These gender-specific norms limit women's access to information by constraining their access to education, their mobility and their interaction with members of the opposite sex. They also limit women's ability to make use of the information that is available to them (Primo 2003).

Rural women's access to financial services

Designing appropriate financial products for women to be able to save, borrow and insure is essential to strengthen women's role as producers and widen the economic opportunities available to them. It is essential to understand how women's access to and control over other resources, including income, shape their need for capital and their ability to obtain it. Farmers and livestock keepers who have access to credit, savings and insurance services can afford to finance the inputs, labour and equipment they need to generate income; they can invest in more profitable enterprises and are more likely to participate in markets more effectively; and can adopt more efficient strategies to stabilize their food consumption meaning they are more food secure (Zeller *et al.* 1997).

Rural women's mobility is often more restricted than men's, which has consequences for their ability to engage in formal financial activities. In some cases women may also be unable to get away from their domestic responsibilities, or may be unable to afford the costs of travel – even when men in households of the same socio-economic level can afford travel (Primo 2003). In any case, women's ability to acquire information will be constrained if, in order to access information, they are required to visit institutions that have inconvenient business hours or are located far from the areas women tend to frequent.

Financial services are often not designed with women in mind. Well-designed products that enable women to adequately save, borrow and insure against unexpected shocks are critical for women to be effective in their production and economic activities. Lending organizations can also be biased against women as their businesses tend to be smaller, more informal, and they lack the necessary collateral (Fletschner 2009).

There are, however, technological innovations and institutional innovations that are making it easier for women to access credit, control their savings and overcome some of the constraints they face (Fletschner and Kenney 2011). These include prepaid cards to distribute loan payments and mobile phone plans to make loan payments and transfer cash. These make it easier for women to sidestep social constraints around mobility (Duncombe and Boateng 2009). The extent to which financial institutions provide both women and men access to and control over individual accounts without the spouse's permission is likely to have a differential impact on men and women's savings rate. For example, Bangladeshi women are constrained from saving large sums of cash since this is likely to attract the attention of male household members who can then take control of those savings. In these circumstances, women are more likely to save only small quantities (Goetz and Gupta 1996). Products such as biometric smart cards allow women to have control over their accounts (Quisumbing and Pandolfelli 2009).

The accumulation of savings is important for enabling women to participate in markets and invest in assets. Because the options and constraints that women face in developing economies differ from those of men, their saving behaviour may also differ. Financial market conditions also interact with gender norms in influencing an individual's saving behaviour. Women's access to and control over income can affect saving behaviour in other ways. Papanek and Schwede (1988), in a Jakarta study, show that women are more likely to participate in *arisan*, informal saving groups, if they are employed. Further, increases in women's earnings raise the household's income and can lead to an increase in saving once basic necessities are met. Equally important, higher relative income improves women's ability to influence the amount of saving out of household income since their fallback position, and thus bargaining power, improves.

A common approach used to increase women's savings are the Village Savings and Loans Associations (VSLAs), and Rotating Savings and Loans Associations (ROSCAs). The VSLAs are a model to create groups of people who can pool their savings in order to have a source of lending funds. Members make savings contributions to the pool, and can also borrow from it. As a self-sustainable and self-replicating mechanism, VSLAs have the potential to bring access to more remote areas.

Analysing men and women's access to information

To analyse gender disparity in access to livestock information within male-headed households, we use data from Kenya. The data was analysed both at household level and compared across men and women within households for different variables of interest, including access to livestock information and financial services. Exploratory analysis of the data was carried out used descriptive and analytical procedures in SPSS and STATA. The exploratory analysis revealed patterns on access to livestock information and financial services by men and women. Further analysis was carried out using the probit model to identify the factors that influenced whether women saved their money. The probit model was based on a dummy dependent variable (1 = women in the household saved money and 0 = they did not save) and a number explanatory variables.

The probit model took the following form:

$$Pr(Y = 1 \mid X) = \emptyset(X'\beta)$$

Where

Pr denotes the probability of women saving or not saving (1 or 0).
X is a vector of regressors on the spouse's and household characteristics.
$\emptyset$ is the Cumulative Distribution Function (CDF) of the standard normal distribution.
β is a parameter typically estimated by maximum likelihood.

Main sources of information on livestock production and marketing

For production-related information, the most common source of information was other farmers, as shown in Figure 6.1. A higher proportion of women (37.6 per cent) than men (32.2 per cent) obtained production information from other farmers. This was also the most common source of marketing information, with 58.4 per cent and 50.4 per cent of women and men respectively obtaining livestock marketing information from other farmers. Groups, associations or cooperatives were the second most important source of information for both men and women, with more women than men getting information from groups on both cattle production and marketing. The third source of information was the radio, which was a source of production information for 11.4 per cent of the women and 14.1 per cent of the men respectively.

Information from government sources was quite low for both men and women and, in some few cases, the farmers accessed information during open days. These findings concur with Brockhaus's (1996) study, which indicated that only 15 per cent of women in southern Jordan were found to have access to state extension services. A higher proportion of women obtained marketing information from other farmers compared to production information.

Similarly, the main source of information for sheep/goats (shoats) production and marketing was from other farmers for both men and women (see Figure 6.2). A higher proportion of men than women (64.9 per cent compared to 53.3 per cent obtained information on sheep and goat marketing from other farmers. For production information, more women than men used other farmers as a source of information. The next common source of information on sheep and goat production and marketing was groups and cooperatives, followed by the radio.

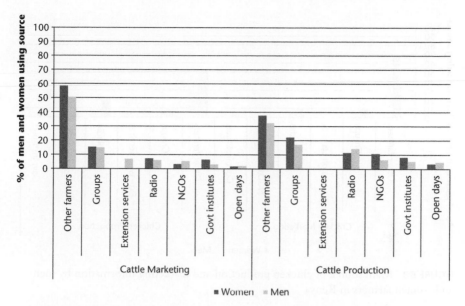

FIGURE 6.1 Main source of cattle production and marketing information by men and women farmers in Kenya

Only 8.2 per cent of women and 5.4 per cent of men obtained information on sheep and goat marketing from the radio, although the percentages were higher for production information at 10.9 per cent and 10.2 per cent for women and men respectively.

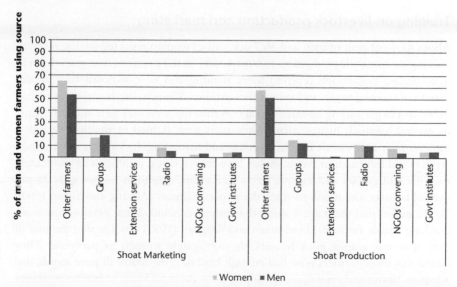

FIGURE 6.2 Main source of sheep and goat production and marketing information by men and women farmers in Kenya

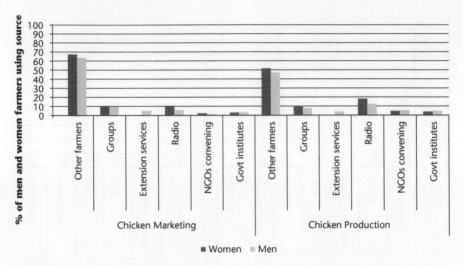

FIGURE 6.3 Main source of chicken production and marketing information by men and women farmers in Kenya

Unlike for cattle, sheep and goats, radio was the second most common source of information on chicken production after other farmers. Both men and women, however, continued to rely on other farmers and their groups for information on marketing. NGOs and government institutes played a very minor role as sources of information, while no men or women farmers received information on chickens from the government extension services.

Training on livestock production and marketing

About 41.4 per cent of men and 36.7 per cent of women reported having received training on livestock production and marketing in the previous five years. Men in male-headed households received more training and were exposed to a greater variety of training topics and venues than women. Women, on the other hand, had access to a larger variety of extension agents than the men, and were trained mainly in general livestock management, while men were trained in multiple technical subjects such as livestock health, breeding and marketing. In looking at extension services and information access, studies have shown that it is difficult to disentangle the effects of gender and income levels. In Zambia, extension reaches only 25 per cent of farmers, and it fails to reach the poorest farmers (Alwang and Siegel 1994). To the extent that these are women, the authors concluded that extension was not reaching female farmers. Hirschmann and Vaughan (1983) observe that the bias of extension was against poor households, not against women in particular. They found that those farmers who had enough land to grow maize in pure stands, had adequate labour and capital, and use inputs were the most likely to receive assistance from extension agents. Because women are under-represented in this group, they were often less likely to obtain assistance.

Some efforts to reach women through extension services have been successful. In Zimbabwe, emphasis has been placed on having extension work with groups, and indeed, women there constitute the majority membership in such groups (Muchena 1994). These groups provide extension services and also make it easier for the women to gain access to credit. Yet women's participation is still constrained by a variety of practices, including the expectation that a woman's husband must approve any legal transaction in which she is involved. Utilization of information may depend on education and literacy levels. Lack of education and higher levels of illiteracy among women farmers may be an additional constraint to women receiving adequate information (Baser 1988).

Only men were trained in marketing. More women than men were trained in general livestock management, processing of products and crop production. Conversely, more men than women were trained in livestock health and livestock breeding, the more technical livestock subjects. Studies have shown that, compared to women, men have easier access to technology and training, mainly due to their strong position as head of the household and greater access to off-farm mobility (Bravo-Baumann 2000). In most countries, research and planning activities in the livestock sector, such as breeding, handling, feeding and health care, are largely dominated by men. Official livestock services are often controlled by men, and extension personnel are primarily men who are not accustomed or trained to teach technical subjects to women. In order to increase the benefits from training, services should be oriented towards those household members who execute these tasks. For example, in societies where sick animals are mainly treated by women, they have knowledge of the symptoms and cures for animal diseases. But with no access to training, progress in best practices and appropriate herding to reduce diseases is difficult. Therefore, where extension services are dominated by men and where women have little access to training due to socio-culturally

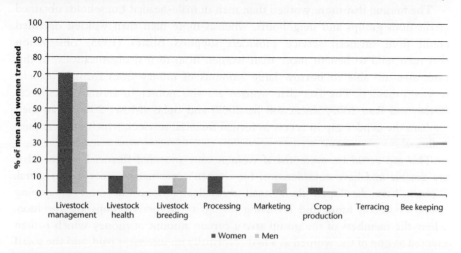

FIGURE 6.4 Percentage of men and women who received training on different crop and livestock practices

defined gender roles, men need to be persuaded to see the relevance and the benefit of training women. Only through a carefully planned gender approach can livestock production goals and successful training of women and men be achieved (Bravo-Baumann 2000).

Training for both men and women was mostly held within the village but outside their homes. Very few men and women were trained in their homes. Increasing access to training by women will require holding training in venues that do not constrain women. The variation in number of men and women trained from home could be because most extension officers are men and are more comfortable talking to men (Shicai and Jie 2009). Gendered disparities in access to training could be overcome if gender roles, relations and ideologies were studied before and during interventions, so that the polarized attitudes and values of men and women are addressed in a way that more women could get involved (Kristjanson *et al.* 2010).

Access to financial services

Access to credit

About 33 per cent of households had obtained cash credit in the five years prior to the survey, as shown in Figure 6.5. For both men and women, groups were the main source of credit. More women received credit from groups and neighbours than men. Men borrowed more from formal credit providers, such as banks and cooperative societies, than women. Although more women (31.5 per cent) than men (28.7 per cent) had received credit, on average, men obtained over three times as much credit (Ksh 60,064 equivalent to US$784.74) as women (Ksh 14,289 equivalent to US$186.69).

The finding that more women than men in male-headed households obtained credit from groups and neighbours, whereas more men than women obtained credit from financial service providers supports Bhatt's (1995) observation that men tend to benefit more than women from formal organizations. In this case, men are able to borrow large amounts of money from formal financial service providers. The fact that women are also poorer in terms of resources (Galab and Rao 2003; Shicai and Jie 2009) and rights (Moser 2006) than men explains the gender gap in access to formal financial services, which often require collateral.

This gap could be overcome if women were provided financial services that are flexible and have consideration for women's constrained access to collateral. Women have developed their small credit/loan systems in most developing countries. Credit funds and revolving savings of women's groups are common, where the members of the group save a certain amount of money which is then granted to one of the women as a loan. Normally no interest is paid, and the social control guarantees that loans are repaid. Other credit systems consist of loans of animals or even milk for processing. Generally, these systems only function at the

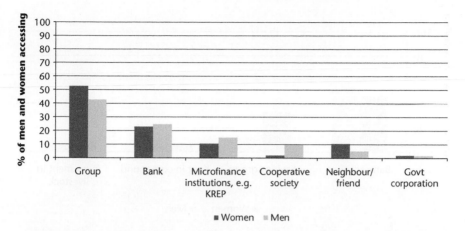

FIGURE 6.5 Percentage of men and women accessing credit from different sources

village level, often between neighbours, where social control can be assured (Bravo–Baumann 2000).

Use of credit

Men used most of their credit on purchase of assets, whereas women spent it on school fees as shown in Figure 6.6. About 19 per cent of both men and women used credit obtained to purchase livestock. Considerably more women (15.8 per cent) spent credit on food purchases than men (1.6 per cent). More than twice the number of women than men borrowed money for construction.

The investments in livestock especially by women support findings in chapter 3 on the main sources of livestock for women, where purchase was the most common means of acquisition of livestock by women.

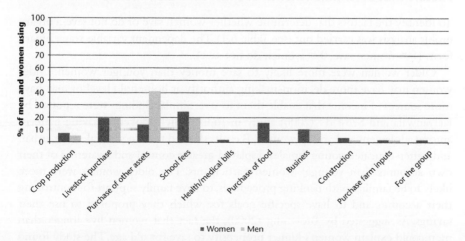

FIGURE 6.6 Different uses of credit by men and women in male-headed households

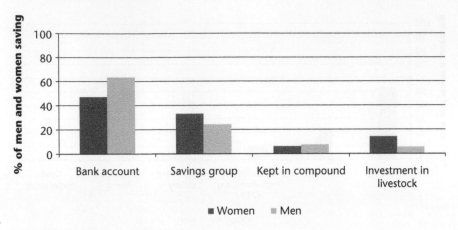

FIGURE 6.7 Percentage of men and women saving money in Kenya

Access to savings

Over 50 per cent of both male- and female-headed households saved money. More men (63.2 per cent) than women (47 per cent per cent) saved money through formal banks (see Figure 6.7). More women than men invested in saving groups and livestock, both of which represent the informal savings mechanisms. These findings confirm the study by Ellis *et al.* (2010) that found that in Kenya men are much more likely to use formal financial services than women (32 per cent of men compared with 19 per cent of women), and women are more likely to use semi-formal services than men (63 per cent of women compared with 58 per cent of men).

Factors that determine whether women save their money

To identify the factors that determine whether women save or do not save, a binary probit analysis was carried out (see Table 6.1). The dependent variable was a binary form (1 = women save, 0 = women do not save).

Older women were more likely to save money than younger women. Older women may have more decision-making authority at household level or may have more sources of income that enable them to save. Similar findings were reported by Kalyanwala and Sebstad's (2006) study in India, which looked at saving patterns among adolescent and young women. Results showed that, in general, older, urban and better-educated young females displayed greater control and awareness of their own accounts than younger women participants. The older women were more likely to be familiar with banking procedures, to have family support for controlling their accounts and to have specific goals for which they proposed to use their savings. As suggested by Browning (2000), the fact that women live longer than men could explain women's higher propensity to save for old age. The study found that the need to save for retirement is also corroborated by the positive and

TABLE 6.1 Factors that determine whether women save their money

Women save (1 = yes, 0 = no)	Coefficient	z	P>z
Age of spouse	0.02	2.26	**0.024**
Primary education(1 = yes)	0.625	2.05	**0.04**
Above primary (1 = yes)	0.917	2.48	**0.013**
Belong to group (1 = yes)	−0.024	−0.5	0.617
Other assets	−0.008	−1.53	0.126
TLU livestock (women)	0.02	0.09	0.925
dist1 = Kajiado	1.019	2.7	**0.007**
dist3 = Meru	−0.388	−1.31	0.191
dist4 = Tharaka	−0.312	−0.91	0.365
Constant	−1.362	−2.2	0.028
Number of observations	172		
LR chi2(9)	37.03		
Prob > chi2	0.0001		
Pseudo R2	0.56		
Log likelihood = −99.960			

diminishing effect of age on the probability of saving. The estimated marginal effect of age implied that a 50-year-old woman is 9 per cent more likely to save than a 40-year woman.

Education was also an important determinant of savings. The more educated women were, the higher the probability they would save. Education empowers women to secure jobs or engage in high income-generating activities enabling them to save their money. The results show that women with primary or above primary-level education were more likely to save compared to women with no education. Increased literacy skills can give women confidence and knowledge of how to engage with formal financial institutions. Browning (2000) found that an extra year of schooling increases the probability of saving by 0.4 per cent and women from households in the highest income quartile are 3 per cent more likely to save. The study further showed that education affects savings performance by influencing the level of income and the options for asset accumulation available to the individual.

The women in Kajiado were more likely to save money compared to women in Kiambu, while the probability of women in Meru and Tharaka saving compared to women in Kiambu was lower, although not significant. These patterns may be related to access to urban centres, with both Kiambu and Kajiado being more urbanized with higher densities of banking services compared to Meru and Tharaka. Rosenzweig (2001) shows that the proximity of formal financial institutions increases financial savings and crowds out informal arrangements. Geographic distance to the nearest bank, or the density of branches relative to the population, can provide a first crude indication of geographic access or lack of physical barriers to access to financial services (Beck and Brown 2011).

Livestock and other assets owned by the women from the sample were found to be insignificant in determining whether women have a way of saving either through formal or informal mechanisms. This was surprising because ownership of assets has often been associated with women's empowerment. Asset ownership influences the "fallback" position of each spouse in negotiations over key household and family decisions, and hence the exit options available to each (Quisumbing and Hallman 2006). In Colombia, Friedemann-Sánchez (2006) found that women use property and social assets to negotiate for the right to work, control their own income, move freely and live without spousal violence. Women's asset ownership may increase the anthropometric status of children (Duflo 2000), the incidence of prenatal care and children's schooling (Doss 2006); it may also reduce domestic violence (Srinivasan and Bedi 2007). Because of these social welfare effects, it is important to have individual-level information on assets in order to find ways to assist women's acquisition of and control over key assets.

Conclusions

Informal channels such as farmer-to-farmer interactions were the key sources of information for livestock production and marketing in the study sites. Information from formal sources such as government extension services was, however, quite limited. Information empowers households in the use of improved technologies and market access, and this can be achieved more through private and public partnerships. More men in male-headed households received more training and were exposed to greater numbers of and more varied topics than women. For women, the training was mainly on general livestock management, mainly done either at home or outside home but within the village. Increasing access to training by women will require holding training in venues that do not constrain women.

About a third of the households interviewed had obtained credit, with groups being the main sources of credit. Men borrowed more from formal credit providers such as banks and cooperatives, while women mainly borrowed from groups and neighbours. This implies that provision of credit facilities should be flexible and have consideration for women's constrained access to collateral. To a considerable degree, women spent more credit on purchase of food than men.

Half of the households surveyed saved their money, with men saving more than women in the formal saving channels, such as banks and cooperatives. Women mainly saved through informal channels, such as groups and in livestock. The provision of accessible and cost-effective financial services is important for smoothing household consumption and the accumulation of incomes and assets.

Probit analysis results on the determinants of savings by women revealed that women's age and education positively and significantly increase the probability of them saving. This implies that older and/or more educated women may have more income, perhaps due to improved job security and earning higher incomes, or they may be more disciplined to save than young women. Systematically targeting older women in micro-credit campaigns could therefore have a positive influence on household-level savings and welfare outcomes such as food security and education. Site-specific analysis shows that access to urban centres increases the probability that women would save money.

References

Alwang J and Siegel P 1994 *Rural Poverty in Zambia: An Analysis of Causes and Policy Recommendations.* Washington, DC: Human Resources Division, Southern Africa Department, The World Bank.

Azzam A M and Schroeter J R 1995 The tradeoff between oligopsony power and cost efficiency in horizontal consolidation: an example from beef packing. *American Journal of Agricultural Economics* 77(4): 825–836.

Baser H 1988 *Technology, Women, and Farming Systems.* Ottawa: Agriculture Canada.

Beck T and Brown M 2011 *Use of Banking Services in Emerging Markets: Household-level Evidence.* CEPR Discussion Papers 8475, Tilburg University, The Netherlands.

Bhatt R E 1995 Women in dairying in India. *Indian Dairyman* 39(2): 157–162.

Bravo-Baumann H 2000 *Gender and Livestock: Capitalisation of Experiences on Livestock Projects and Gender.* Working document. Swiss Agency for Development and Cooperation, Bern.

Brockhaus M 1996 *The Role of Women on Sheep and Goat Farms in Jordan.* Amman, Jordan: GTZ.

Browning M 2000 The saving behaviour of a two-person household. *Scandinavian Journal of Economics* 102(2): 235–251.

Chowdhury S 2006 Access to a telephone and factor market participation of rural households in Bangladesh. *Journal of Agricultural Economics* 57(3): 563–576.

De Haan N 2001 Of goats and groups: a study on social capital in development projects. *Agriculture and Human Values* 18(1): 71–84.

Dercon S and Hoddinott J 2005 *Livelihoods, Growth, and Links to Market Towns in 15 Ethiopian Villages.* FCND Discussion Paper 194. Washington, DC: IFPRI.

Doss C R 2006 The effects of intra household property ownership on expenditure patterns in Ghana. *Journal of African Economies* 15(1): 149–180.

Duflo E 2000 *Grandmothers and Granddaughters: Old-Age Pension and Intrahousehold Allocation in South Africa.* Working Paper Series No. 8061. Cambridge, MA: National Bureau of Economic Research.

Duncombe R and Boateng R 2009 Mobile phones and financial services in developing countries: a review of concepts, methods, issues, evidence and future research directions. *Third World Quarterly* 30(7): 1235–1258.

Ellis B, Lemma A and Rud J 2010 Investigating the impact of access to financial services on household investment. *Chart August.*

Fletschner D 2009 Rural women's access to credit: market imperfections and intrahousehold dynamics. *World Development* 37(3): 618–631.

Fletschner D and Kenney L 2011 *Rural Women's Access to Financial Services – Credit, Savings and Insurance.* ESA Working Paper No. 11. Rome: FAO.

Friedemann-Sánchez G 2006 Assets in intrahousehold bargaining among women workers in Colombia's cut-flower industry. *Feminist Economics* 12(1–2): 247–269.

Galab S and Rao C 2003 Women self-help groups: poverty alleviation and empowerment. *Economic and Poverty Weekly* 38(12/13): 1274–1283.

Goetz A M and Gupta R S 1996 Who takes the credit? Gender, power, and control over loan use in rural credit programs in Bangladesh. *World Development* 24(1): 45–63.

Hirschmann D and Vaughan M 1983 Food production and income generation in a matrilineal society: rural women in Zomba, Malawi. *Journal of Southern African Studies* 10(1): 86–99.

IDRC 1998 *Telecentre Research Framework for ACACIA*. Canada: IDRC.

IFAD 2004 *Livestock Services and the Poor: A Global Initiative. Collecting, Coordinating and Sharing Experiences*. Rome: IFAD.

Kalyanwala S and Sebstad J 2006 *Spending, Saving and Borrowing: Perceptions and Experiences of Girls in Gujarat*. New Delhi: Population Council.

Kizito A M 2011 *The Structure, Conduct, and Performance of Agricultural Market Information Systems in Sub-Saharan Africa*. PhD dissertation, Michigan State University.

Kristjanson P, Johnson N, Tipilda A, Njuki J, Baltenweck I, Grace D *et al.* 2010 *Livestock and Women's Livelihoods: A Review of the Recent Evidence*. ILRI Discussion Paper No. 20. Nairobi: ILRI.

Moser C 2006 *Asset-based Approaches to Poverty Reduction in a Globalized Context: An Introduction to Asset Accumulation Policy and Summary of Workshop Findings*. Washington, DC: Brookings Institution.

Muchena O N 1994 The changing perceptions of women in agriculture. In Eicher M R and Eicher C K (eds) *Zimbabwe's Agricultural Revolution*. Harare: University of Zimbabwe Press.

Munyua H 2000 Information and communication technologies for rural development and food security: lessons from field experiences in developing countries. Workshop paper, Sustainable Development Department (SD), FAO. Rome: FAO. Available at: http://www.fao.org/sd/cddirect/CDre0055b.htm (accessed May 2013).

Papanek H and Schwede L 1988 Women are good with money: earning and managing in an Indonesian city. In Dwyer D and Bruce J (eds) *A Home Divided: Women and Income in the Third World*. Stanford, CA: Stanford University Press, pp. 71–98.

Primo N 2003 *Gender Issues in the Information Society*. Paris: UNESCO Publications for the World Summit on the Information Society.

Quisumbing A and Pandolfelli L 2009 *Promising Approaches to Address the Needs of Poor Female Farmers: Resources, Constraints, and Interventions*. IFPRI Discussion Paper No. 882, Washington, DC: IFPRI.

Quisumbing A and Hallman K 2006 Marriage in transition: evidence on age, education, and assets from six developing countries. In Lloyd C B, Behrman J R, Stromquist N P and Cohen B (eds) *The Changing Transitions to Adulthood in Developing Countries: Selected Studies*. Washington, DC: National Academies Press.

Romer P 1993 Ideas gaps and object gaps in economic development. *Journal of Monetary Economics* 32(3): 543–573.

Rosenzweig M R 2001 Savings behaviour in low-income countries. *Oxford Review of Economic Policy* 1: 40–54.

Shicai S and Jie Q 2009 Livestock projects in southwest China: women participate, everybody benefits. *Leisa Magazine* 25(3 Sept.).

Srinivasan S and Bedi A 2007 Domestic violence and dowry: evidence from a south Indian village. *World Development* 35(5): 857–880.

Stango V and Zinman J 2008 *The Price is Not Right (Not Even on Average): Exponential Growth Bias, Present-biased Perceptions and Household Finance*. Working paper, Dartmouth College. Hanover, NH: Dartmouth College.

Zeller M, Schrieder G, von Braun J and Heidhues F 1997 *Rural Finance for Food Security for the Poor*. Washington, DC: IFPRI.

7

WOMEN, LIVESTOCK OWNERSHIP AND FOOD SECURITY

Juliet Kariuki, Jemimah Njuki, Samuel Mburu and Elizabeth Waithanji

Introduction

At the World Food Summit in 1996, food security was defined as when all people, at all times, have physical and economic access to sufficient safe and nutritious food that meets their dietary needs and food preferences for an active and healthy life. There are several dimensions to food security: availability, access, utilization and stability (FAO 2008; Pinstrup-Anderson 2009). There are several causes of food insecurity, including low purchasing power due to low incomes and poverty; low food production caused by low productivity, drought and other factors; and poor food distribution systems (Gladwin *et al.* 2001; Uvin 1994). There are also gender dimensions to food insecurity, the most studied of which is the role that men and women play in food production, processing and distribution (Dey 1984; Gittinger *et al.* 1990; Rengam 2001). Less studied, however, is the role that gender inequalities (in resource allocation, income management, access to productive resources) play in causing food insecurity. The Food and Agriculture Organization (FAO) report on women and agriculture (FAO 2011) indicates that closing the gender gap in access to productive resources could increase agricultural output by 2.5–4 per cent and reduce the number of undernourished people by 12–17 per cent. Kennedy and Peters (1992) found an interaction between the total household income and proportion of income controlled by women and a household's caloric intake. The FAO report (FAO 2011) indicates that when women control additional income, they spend more of it than men do on food, health, clothing and education of children.

Livestock plays an important role in contributing to food security, through: (i) enabling direct access to animal source foods; (ii) providing cash income from sale of livestock and livestock products that can in turn be used to purchase food especially during times of food deficit; (iii) contributing to increased aggregate cereal supply as a result of improved productivity from use of manure and traction; and (iv) lowering prices of livestock products and, therefore, increasing access to such products by the poor, especially poor urban consumers through increasing livestock production. While analyses of the role of livestock-keeping in influencing consumption of animal source foods have been done, little work on the role that livestock play in buffering households against food deficit, as well as the implications of women's ownership of livestock in influencing food security has been conducted. A deficiency of analysis on the influence of intra-household livestock ownership patterns and decision-making on household food security limits the extent to which livestock can effectively be used as an intervention for improving food security.

This chapter uses three indicators of food security to analyse the links between household and intra-household livestock ownership and food security. The indicators used are the household dietary diversity scores (HDDS), the months of adequate household food provisioning (MAHFP) and the frequency of consumption of animal source foods. The HDDS measures the number of different food groups consumed over a given reference period. It is an attractive indicator as a more diversified diet is an important outcome in and of itself, and is associated with a number of improved outcomes in areas such as birth weight, child anthropometric and nutritional status (Arimond and Ruel 2004; Steyn *et al.* 2006), and improved haemoglobin concentrations. A more diversified diet is highly correlated with such factors as caloric and protein adequacy, percentage of high-quality protein from animal sources and household income. Even in very poor households, increased food expenditure resulting from additional income is associated with increased quantity and quality of the diet (Low 1991, cited in Nicholson and Thornton 1999). Questions on dietary diversity can be asked at the household or individual level, making it possible to examine food security at the household and intra-household levels.

The MAHFP is an indicator of food access which depends on the ability of households to obtain food from their own production, stocks, purchases, gathering, or through food transfers from relatives, members of the community, the government or donors. A household's access to food also depends on the resources available to individual household members and the steps they must take to obtain those resources, particularly exchange of other goods and services. Measuring the MAHFP has the advantage of capturing the combined effects of a range of interventions and strategies, such as improved agricultural production, storage and interventions that increase the household's purchasing power.

The frequency of consumption of animal source foods is based on the food consumption score (FCS), which is a comprehensive indicator based on dietary diversity, food frequency and relative nutritional importance (WFP 2008). The FCS is a flexible and efficient approach to the collection of food consumption data that can be used to capture important food security indicators such as dietary diversity and food frequency in poor rural and urban households.

Links between livestock production, gender and food security

Livestock contributes to food security in multiple ways as shown in Figure 7.1. The role of animal source foods (ASFs) in the reduction of micro nutrient deficiencies and enhancement of dietary adequacy is widely accepted (Leroy and Frongillo 2007; Murphy and Allen 2003). ASFs such as milk, meat and eggs are rich in energy and also provide a good source of proteins, vitamins and minerals. The nutrients derived from ASFs are more easily absorbed than the same nutrients found in plant source foods, which are often consumed in the form of rice and maize by poor households (Arimond and Ruel 2004; Faber 2010). In developing countries, the poor consumption of micro nutrients found in ASFs can lead to inadequate nutritional status and can contribute to increased mortality rates (Black *et al.* 2008). The consumption of ASFs, particularly for the rural poor, can therefore contribute substantially to dietary diversity and household nutritional status and, as a consequence, has implications for household productivity, income levels (Leroy and Frongillo 2007) and ultimately national development (FAO 2000, in Speedy 2003; Sanghvi 1996, in Welch and Graham 2000). Studies on the role of livestock production in contributing to the consumption of ASFs have often overlooked the role of intra-household allocation of resources as well as preferences.

Another important component of livestock as a contributor to household food security is as a buffer during times of food shortage (Kinsey *et al.* 1998, cited in Fisher *et al.* 2010). Sales of livestock and livestock products provide purchasing

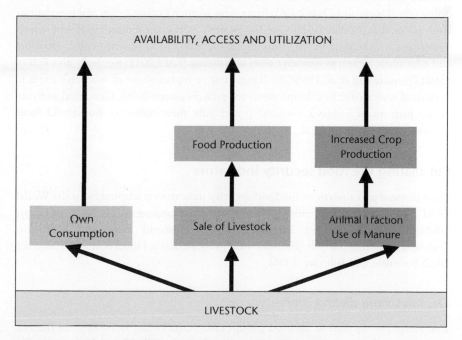

FIGURE 7.1 Relationship between livestock and food security

power, and thus access to food. In many cases, the sale of livestock is the only outlet of smallholders in rural communities to the monetary economy. In West Africa, for example, Fafchamps *et al.* (1998) estimated that livestock sales compensated for at most 30 per cent, and probably closer to 15 per cent of income shortfalls due to village-level shocks alone, while in Burkina Faso Reardon *et al.* (1988) found that livestock and stored grain were among the strategies that households used to address food shortages in the event of a crop failure. Even in times of food abundance, livestock sales can enable households to diversify their diets (Fratkin and Smith 1995). Generally the degree of commercialization of livestock products is much higher than that of crops in developing countries. For example Ehui *et al.* (1998) report that in areas of extreme poverty, such as the Central Highlands of Ethiopia, the sale of such products as dung cake is the most important source of cash for households. In semi-arid Mali, livestock contributed 78 per cent of cash income from crops and livestock in smallholder mixed farming (Debrah and Sissoko 1990), while in pastoral areas of East Africa, sale of livestock and milk is the main source of income for the purchase of grain for household consumption (Little 1996).

The extent to which livestock contributes to food security is, however, dependent on intra-household dynamics, such as women's ownership of assets and the extent to which they can make decisions on the use of the assets, and on how much of the products will be sold or will be used for home consumption. The intra-household allocation of assets has important implications for a range of outcomes. A growing body of empirical evidence has shown that not only do women typically have fewer assets than men, but they also use the ones they have differently (Deere and Doss 2006). Increasing women's control over assets – mainly land, physical and financial assets – has positive effects on a number of important development outcomes for the household, including food security, child nutrition and education, as well as women's own well-being (FAO 2011; Kennedy and Peters 1992; Quisumbing *et al.* 1995). In Bangladesh, a higher share of women's assets is associated with better health outcomes for girls (Hallman 2000). Gendered analyses of the patterns of livestock ownership and how these influence household food security are therefore critical.

Calculating the food security indicators

Measurement and analysis of the food security indicators is adapted from the World Food Programme's (WFP) vulnerability assessment mapping (WFP 2008) and from USAID's (US Agency for International Development) Food and Nutrition Technical Assistance Project (Bilinsky and Swindale 2010; Hoddinott and Yohannes 2002; Swindale and Bilinsky 2006).

The household dietary diversity score

The HDDS is the sum of all food groups consumed by the household in the last 24 hours divided by the total number of food groups. The dietary diversity score should ideally be measured at individual household member level. However, due to both time and budget constraints questions were asked only of the female spouses as, in most instances, they are responsible for, and most aware of the food consumed within the household. For the measurement of the HDDS, the different types of foods are grouped into 12 categories as shown in Table 7.1.

The HDDS ranges between 0 and 1. A value of 1 signifies that a household consumed all the 12 food groups in the last 24 hours. An increase in the average number of different food groups consumed provides a quantifiable measure of improved household food access. In general, any increase in HDDS reflects an improvement in the household's diet. While the dietary diversity score offers an accepted and popularized proxy for calorie intake and nutritional outcomes (Ruel 2003), this approach still has some limitations. For example, seasonal variations are obscured if dietary diversity data is not collected at strategic intervals during the year (Savy *et al.* 2006). Poor precision of dietary diversity measurements may also limit the accuracy of data.

Calculating the food consumption score

To calculate the FCS, the 12 categories of food are reduced to 9 main food groups based on their contribution to diets (see Table 7.2); main staples (A and B), vegetables (C), fruits (D), pulses (E), meat, poultry, fish and eggs (F, G, H, I), milk (J), oil (K), sugar and condiments and drinks (L). The food types are weighted based on nutrient densities estimated by WFP for use in vulnerability assessment and mapping (WFP 2008). The FCS is calculated by first working out the consumption frequencies

TABLE 7.1 Food categories used for calculating the HDDS

	Description
A	Any foods including local foods made from starch foods, e.g. bread, rice, noodles, biscuits or any other foods made from millet, sorghum, maize, rice, wheat or other local grains
B	Potatoes, yams, cassava, or any other foods made from roots or tubers
C	Vegetables, including local green leafy vegetables
D	Fruits
E	Beef, pork, lamb, goat, rabbit, wild game, duck or other birds; liver, kidney, heart and other organ meats
F	Eggs
G	Fresh or dried fish, shell fish
H	Beans made from grain legumes such as beans, peas, lentils or nuts
I	Milk and milk products including cheese, yoghurt and other local dairy products
J	Foods made with oil, fat or butter
K	Sugar and sugar products, honey
L	Condiments and drinks such as coffee, tea

TABLE 7.2 Food groups used for the calculation of the FCS and their assigned weight

Types of foods	Groups	Weight
A. Staples or food made from staples including millet, sorghum, maize, rice, wheat, or other local grains, e.g. ugali, bread, rice noodles, biscuits, or other foods	**Main Staples** (if sum of frequencies is > 7 set to 7)	2
B. Potatoes, yams, cassava or any other foods made from roots or tubers		
C. Vegetables	**Vegetables**	1
D. Fruits	**Fruits**	
E. Beans, peas, lentils, or nuts	**Pulses**	3
F. Red meat: beef, pork, lamb, goat, rabbit, wild game, liver, kidney, heart, or other organ meats	**Meat and Fish** (if sum of frequencies is > 7 set to 7)	4
G. Poultry, including chickens, duck, other poultry		
H. Eggs		
I. Fresh or dried fish or shellfish		
J. Milk, cheese, yoghurt, or other milk product	**Milk**	4
K. Oils and fats	**Oil**	0.5
L. Sweets, sugar, honey	**Sugar**	0.5
M. Any other foods, such as condiments, and drinks such as coffee, tea including milk in tea	**Condiments**	0

(number of times the food type was eaten in the last seven days) for each food group. For food groups that combine different types of food, the frequencies from each food type are summed up to provide a total for the food group. The maximum frequency for a food group is 7, so if the total frequency amounts to greater than 7 then this is replaced with 7. The final step is a multiplication of the frequency of each food group by its weight and an addition of the weighted food group scores to create the FCS. Thresholds can be determined based on the consumption behaviour of the country or region under consideration. The WFP, for example, uses the following thresholds: an FCS of 0–21 is poor, 21.5–35 is borderline, while above 35 is good.

Calculating the MAHFP

This is calculated by adding all the months that a household had adequate food in the previous 12 months. An average for the sample may be obtained by adding all the MAHFP and dividing by the number of households. The denominator should include all households interviewed, including those that did not experience any months of food shortage. The MAHFP indicator ranges from 0 to 12 months. A value of 0 indicates that a household did not have enough food in any month during the last 12 months while a value of 12 shows that the household had enough food during all the 12 months. The indicator currently does not have thresholds but households can be classified as belonging to the top, middle and lower tercile (Bilinsky and Swindale 2010).

Livestock ownership, dietary diversity and food adequacy

Numerous studies illustrate the contribution of livestock production to food security; however, little empirical evidence shows the relationship between ownership of different livestock species and the dietary diversity of poor rural households. Results show that households owning livestock had a higher, but not always significantly higher HDDS than non-livestock–owning households. Households owning goats and exotic chickens had significantly higher HDDS than those that did not own these species. These households had two times the HDDS of non-goat-, non-chicken-owning households, implying that they consumed twice as many food categories as households that did not own goats and chickens (Table 7.3).

Results show no significant differences in MAHFP between livestock-owning and non-livestock–owning households. It was hypothesized that in times of food shortage, households would sell their livestock and livestock products to purchase food. It would seem that while they did this to augment their diets and therefore increase the HDDS, livestock sales were not used for bulk food purchase that would reduce the number of months those households did not have enough food.

Although local chicken consumption would seem more common at household level, exotic chickens – sometimes characterized by a higher level of productivity (Alemu and Tadelle 1997, in Bogalle, 2010), commercialization, and often considered a venture of wealthier families (Alemu and Tadelle 1997, in Bogalle, 2010; Alders

TABLE 7.3 Dietary diversity and food adequacy in livestock- and non-livestock–owning households

Mean HDDS by livestock ownership

Livestock	Own livestock	No livestock	t-value
Cattle	0.72	0.33	1.1380
Goats	0.61	0.36	6.165*
Exotic chickens	0.76	0.34	6.813*
Local chickens	0.69	.25	0.643

Mean MAHFP by livestock ownership

Livestock	Own livestock	No livestock	t-value
Cattle	7.2	9.3	1.1570
Goats	6.6	9.1	1.5970
Exotic chickens	7.1	9.3	0.734
Local chickens	6.5	5.3	0.207

*1%, **5%,***10% significance level

et al. 2001) – has a highly lucrative market. Research by Alemu and Tadelle (1997, cited in Bogalle 2010) shows that in some African countries exotic chickens can contribute up to 70 per cent of the total poultry population. It is therefore possible that large income gains from exotic chicken sales enabled households from our sample to purchase a greater diversity of food and enjoy a more diversified diet than households that rely on incomes from traditional chicken sales. Households owning exotic chickens also consumed significantly more eggs than those that did not, whereas consumption of eggs from local chickens did not differ significantly between those households that owned local chickens and those that did not (see Table 7.4).

The nutritional benefits of owning dairy goats and engaging in dairy goat marketing have been recorded in several countries. In East Africa, projects that focus on improving the live dairy goat and goat milk value chain offered rural producers an opportunity to benefit from high-value markets, increases in household-level incomes and improved household nutrition (Peacock 2008). In Ethiopia, a livestock development project improved the nutritional status of women and children by increasing ownership of household dairy goats (Peacock 2005). In China, stakeholders of a similar project hypothesized that improved household food security and nutrition resulted from income generated from the sale of goat kids, meat and hides (Sinn *et al.* 1999). In Ethiopia, a goat development program that aimed to improve access to and consumption of ASFs in rural households found that consumption of both dairy goat and poultry products was closely correlated with increased ownership and productivity of dairy goats (Ayele and Peacock 2003). However, the project only recorded increases in nutrient-rich vegetable consumption in locations where training and community-based nutrition activities were conducted (Ayele and Peacock 2003). In Bangladesh, increased incomes from a smallholder poultry development project resulted in more diversified diets for beneficiaries, including grains, milk and poultry products (Alam 1997). While in Africa much evidence shows a correlation between ownership of village chickens and protein intake in the form of poultry and egg consumption (Kitalyi 1998; Nielsen *et al.* 2003), there have been some recorded exceptions. For example, according to Aklilu *et al.* (2008), in Ethiopia consumption is not the main purpose of keeping poultry because poultry product consumption only partly meets the protein needs of the households in the study.

The consumption of ASFs is an important component of food security and human nutrition. One approach to comprehensively capture the influence of livestock ownership on consumption of ASFs is through the use of the FCS. Results show that ASFs contributed considerably to household food consumption, particularly in livestock-owning households. There was, however, an exception with regard to meat consumption. The frequency of meat consumption was the highest in households with no cattle (3.2), goats (3.3) and exotic chickens (3.3) compared with households that did own these livestock. Households that kept exotic chickens consumed eggs with the highest frequency (2.3); whereas households that did not own local chickens did not consume any of the ASFs reviewed (Table 7. 4).

TABLE 7.4 Number of times per week that households consumed ASFs in livestock- and non-livestock-owning households

Livestock		Own livestock	No livestock	t-value
Number of times households consumed meat if they owned or did not own specific species	Cattle	1.7	3.2	5.036*
	Goats	1.2	3.3	3.585*
	Exotic chickens	3.1	3.3	0.389
	Local chickens	1.4	0	
Number of times households consumed eggs if they owned or did not own specific species	Exotic chickens	2.6	1.9	1.813***
	Local chickens	2.3	0	
Number of times households consumed milk if they owned or did not own specific species	Cattle	6.5	4	7.386*
	Goats	5.6	4.8	1.2

*1%, **5%,***10% significance level

As expected, households with cattle consumed milk and meat more frequently than households that did not have cattle, while ownership of goats also led to a higher frequency of consumption of meat but not milk. Households with exotic chickens also consumed eggs more frequently than those that did not own chickens. There were no significant differences in egg consumption in households that owned and did not own local chickens, most likely due to the low numbers of local chickens kept by farmers and their lower productivity.

Increases in income and livestock production have a positive impact on consumption of ASFs according to numerous studies. For example, the relationship between milk consumption and dairy cow ownership was explored by Nicholson et al. (2003). Results showed that dairy cow ownership increased consumption of dairy products by 1.0 litre per week, whereas, evidence from a smallholder poultry development project in Bangladesh revealed that increased incomes resulted in increased consumption of chicken and eggs (Alam 1997). There is also evidence that describes the influence of livestock ownership on the consumption of meat. Scoones (1992, in Randolph et al. 2007) explored consumption patterns and found that meat consumption from own slaughter was infrequent except in cases of sick and/or unproductive animals or for ceremonial reasons. Results from a goat development program in Ethiopia found that 63 per cent of project beneficiaries slaughtered goat meat for consumption; 37 per cent of this group consumed goat meat during holidays and/or festivities, while 63 per cent slaughtered goats for events such as births and funerals (Ayele and Peacock 2003). According to Faber's (2010) analysis of nutrition in vulnerable communities in economically marginalized societies, three possible constraints limit the consumption of ASFs; availability, affordability and lack of cold storage facilities.

Evidence of the rapid growth in global livestock production is considered a result of increasing demand for animal products. However, statements such as these disguise the fact that increased livestock production is confined only to certain countries or regions, and is not taking place in some of the poorer African countries. According to Speedy (2003), ASF consumption is declining in these countries as population increases. In fact, Mozambique and Kenya are among the countries that consume the least amount of meat and fish in Africa (Speedy 2003). These results can support efforts that seek to improve food production as a means of promoting increased consumption of ASFs and food security for the poor and by the poor.

Ownership of livestock by women and food security

Ownership of livestock by women can influence the decisions they make on how to use that livestock or livestock products, as well as how to use other streams of benefits, for example, income emanating from that livestock. The results on women's ownership of livestock and food security were mixed. The most significant differences in women owning livestock were observed for the MAHFP as shown in Table 7.5. It is likely that women were able to dispose of livestock such as goats and chickens, or had more decision-making authority on disposal of the livestock products to purchase food.

The commercialization of the livestock sector can create a pathway out of poverty for smallholder women livestock keepers. However, for research to show how best to maximize women's market benefits, it is important to establish the dynamics which influence intra-household income control and resource allocation. Indeed, several studies show a positive relationship between increased incomes under the management of women, including improvements in child nutritional status and dietary intake (Bennett 1988; Kumar 1977). It is therefore possible that incomes generated by owners of exotic chickens may have been influenced by similar factors identified in previous studies. However, it may be possible that women's financial status can also serve to subjugate them further, especially if men's household expenditure reduces as women manage more income. For example, a study in Nigeria (Aromolaran 2004) found increases in women's income share slightly reduced per capita calorie intake, which conflicts with the hypothesis that increases in the share of income under women's control will increase calorie intake.

TABLE 7.5 Months of adequate household food provisioning in households where women owned or did not own different livestock species

Livestock	Women own livestock species	Women do not own livestock species	t-value
Cattle	9.2	8.3	−3.602*
Goats	8.9	8.4	−1.639***
Exotic chickens	9.7	8.4	−1.677***
Local chickens	8.7	8.4	−1.621

*1%, **5%,***10% significance level

These results also suggest that the redistribution of intra-household income from male household heads to female spouses, as is sometimes promoted through development interventions and enforced through food security policies, may not yield desirable food calorie intake outcomes (Aromolaran 2004).

The relationship between women's ownership of livestock and the consumption of different ASFs is consistent for some species and products and not others. Results indicate that the frequency of meat consumption in households where women owned livestock was considerably higher than in households where women did not own livestock. There were significant differences in meat consumption in households where women owned exotic chickens (t-value of 2.552) and cattle (t-value of 2.268). The frequency of milk consumption was significantly lower in households where women owned cattle compared with households where women did not own cattle (t-value of 2.281). This may be a general reflection of women's lower decision-making on large livestock such as cattle (see Table 7.6).

When combined, the above results show that if women own livestock, the number of months during which households have adequate food increases, as does the consumption of some ASFs. An emerging body of research shows that women's livestock husbandry and agricultural roles, such as the care and management of dairy goats (Peacock 2005), the ability to derive income from small ruminant sales in Kenya (Oxby 1983, in Ajala 1995) and women's pivotal role in the processing, marketing and storage of agricultural produce can influence their ability to provide food for the household (Ajala 1995). However, women's often limited control over productive assets and income management remains a potential risk to their ability to boost household food security. These findings suggest that interventions intended to improve household nutrition outcomes can face limited success if women and men are not addressed jointly as beneficiaries, and if there is

TABLE 7.6 Consumption of ASFs in households where women own and do not own different livestock species

	Livestock	Number of times eaten per week in households where women own species	Number of times eaten per week in households where women do not own species	t-value
Consumption of meat	Cattle	2.7	1.9	2.268★★
	Goats	3	2.2	1.928
	Exotic chickens	4	3.2	2.552★
	Local chickens	2.4	2.1	0.802
Consumption of eggs	Exotic chickens	2.4	2.5	0.835
	Local chickens	1.9	1.8	0.46
Consumption of milk	Cattle	3.9	6.5	2.281★★
	Goats	4	4	1.1

★★★1%, ★★5%, ★10% significance level

only a narrow investment in the different types of capital necessary for development by the rural poor (Berti *et al.* 2004; Scanlan 2004).

Research shows that livestock ownership at the household level is gendered, with women more likely to own small stock and/or less valuable livestock than men (see chapter 3, this volume). Based on these observations, women in male-headed households from our sample who owned valuable exotic poultry and large stock like cattle may be considered further up the livestock ladder than other women, who owned or did not own local chickens. It is likely that revenues from livestock and livestock product sales under the control of women contributed significantly to household consumption of meat. If so, these results would correspond with a study on poultry in Bangladesh that found that income increases from chicken sales increased the consumption of ASFs (Alam 1997).

In much of East Africa, dairying by smallholder farm families is viewed by governments and development agencies as a means of increasing the production of needed nutrients, and as a source of cash income to purchase other foods (Staal *et al.* 1997, in Nicholson and Thornton 1999). These results, however, indicate that livestock ownership does not always necessarily increase the frequency of consumption of ASFs and the diversity of diets. Despite findings that show that the number of dairy cows owned can significantly impact household cash incomes when compared with households without dairy cows, Nicholson and Thornton (1999) found that dairy cattle ownership does not always translate to an equivalent improvement in nutritional outcomes.

It should be noted that other factors can influence the diversity of diets, food adequacy and consumption of ASFs and these should also be considered when designing food security initiatives. For example, there is evidence to suggest that if the cost of producing livestock products domestically is lower than its production in the commercial sector, households are more likely to opt for the sale of these products, rather than domestic consumption (Jensen and Dolberg 2003). Research from Nigeria shows that, at the individual household level, poor producer families are less inclined to consume poultry products and more likely to sell them, especially when the household is in need of cash (Sonaiya 2009).

Conclusion

Livestock ownership plays a vital role in enabling households to benefit from a more diverse diet, and in contributing to the consumption of ASFs. While this study concurs with others that find positive relationships between livestock ownership and food security, the results show that livestock species and owner-ship patterns play an equally significant role in determining household food consumption. Livestock ownership and intra-household ownership patterns influence different indicators of food security in different ways. Some livestock species are also more important for food security and for consumption of ASFs than others. Assuming that dietary diversity is increased by households' ability to purchase foods that they do not produce, having small stock such as chickens and goats that can easily be sold is much more likely to influence dietary diversity than the larger livestock.

Ownership of livestock by women can increase the probability that women will make decisions on allocation of livestock, livestock products or income derived from these for household consumption, increasing the likelihood that households consume these products. Further analysis is, however, required to look at the intra-household allocation of these foods as well as other factors that may influence consumption of ASFs.

The results illustrate the advantages of utilizing a variety of food security indicators to identify a more comprehensive account of food security status in the households sampled and to further identify which indicators of food security are influenced by livestock ownership and intra-household ownership patterns. And, as Gittinger *et al.* (1990) recommends, in seeking ways to improve household food security in Africa it is important to intervene in ways that women benefit from, such as improving their ownership of assets and enhancing their decision-making abilities while being careful not to increase women's burden of production and household food provisioning.

References

Ajala, A. (1995). Women's tasks in the management of goats in southern Nigeria. *Small Ruminant Research* 15(3): 203–208.

Aklilu, H., Udo, H., Almekinders, C. and Vanderzijpp, A. (2008). How resource poor households value and access poultry: village poultry keeping in Tigray, Ethiopia. *Agricultural Systems* 96(1–3): 175–183.

Alam, J. (1997). Impact of smallholder livestock development project in some selected areas of rural Bangladesh. *Livestock Research for Rural Development* 9(3): 1–14. Available at: http://lrrd.cipav.org.co/lrrd9/3/bang932.htm (accessed 29 July 2013).

Alders, R. G., Spradbrow, P. B., Young, M. P., Mata, B.V., Meers, J., Lobo, Q. J. P. *et al.* (2001). Improving rural livelihoods through sustainable Newcastle disease control in village chickens. In *Proceedings of the 10th International Conference of the Association of Institutions for Tropical Veterinary Medicine*, 20–23 August, Copenhagen, pp. 199–205.

Arimond, M. and Ruel, M.T. (2004). Dietary diversity is associated with child nutritional status: evidence from 11 demographic and health surveys. *Journal of Nutrition* 134(10): 2579–2585.

Aromolaran, A. (2004). Household income, women's income share and food calorie intake in southwestern Nigeria. *Food Policy* 29(5): 507–530.

Ayele, Z. and Peacock, C. (2003). Improving access to and consumption of animal source foods in rural households: the experiences of a women-focused goat development program in the highlands of Ethiopia. *Journal of Nutrition* 133(11): 3981S–3986S.

Bennett, L. (1988). The role of women in income production and intra-household allocation of resources as a determinant of child nutrition and health. *Food and Nutrition Bulletin* 10(3): 16–26.

Berti, P. R., Krasevec, J. and FitzGerald, S. (2004). A review of the effectiveness of agriculture interventions in improving nutrition outcomes. *Public Health Nutrition* 7(5): 599–609.

Bilinsky, P. and Swindale, A. (2010). *Months of Adequate Household Food Provisioning (MAHFP) for Measurement of Household Food Access: Indicator Guide.* Washington DC: Food and Nutritional Technical Assistance Project, Academy for Educational Development.

Black, R. E., Lindsay, H. A., Bhutta, A. Z., Caulfi, L., De Onis, M., Ezzati, M. *et al.* (2008). Maternal and child undernutrition: global and regional exposures and health consequences. *The Lancet* 371(9608): 243–260.

Bogalle, M. M. (2010). Characterization of village chicken production and marketing system in Gomma Wereda, Jimma Zone, Ethiopia. MSc thesis, Jimma University, Ethiopia.

Debrah, S. and Sissoko, K. (1990). Sources and transfers of cash income in the rural economy: the case of smallholder mixed farmers in the semi-arid zone of Mali, African Livestock Policy Analysis Network (ALPAN) No. 25. Addis Ababa, Ethiopia: ILCA.

Deere, C. D. and Doss, C. R. (2006). The gender asset gap: what do we know and why does it matter? *Feminist Economics* 12(1–2): 1–50.

Dey, J. (1984). *Women in Food Production and Food Security in Africa: Women in Agriculture.* Working Paper No. 3. Rome: FAO.

Ehui, S., Li-Pun, H., Mares, V. and Shapiro, B. (1998). The role of livestock in food security and environmental protection. *Outlook on Agriculture* 27(2): 81–87.

Faber, M. (2010). Nutrition in vulnerable communities in economically marginalized societies. *Livestock Science* 130(1–3): 110–114.

Fafchamps, M., Udry, C. R. and Czukas, K. (1998). Drought and saving in West Africa: are livestock a buffer stock? *Journal of Development Economics* 55: 273–305.

FAO (2008). *An Introduction to the Basic Concepts of Food Security.* Rome: EC–FAO Food Security Programme.

FAO (2011). *The State of Food and Agriculture 2010–2011. Women in Agriculture: Closing the Gender Gap for Development.* Rome: FAO.

Fisher, M., Chaudhury, M. and McCusker, B. (2010). Do forests help rural households adapt to climate variability? Evidence from Southern Malawi. *World Development* 38(9): 1241–1250.

Fratkin, E. and Smith, K. (1995). Women's changing economic roles with pastoral sedentarization: varying strategies in alternate Rendille communities. *Human Ecology* 23(4): 433–454.

Gittinger, J., Chernick, S. and Horenstein, N. (1990). *Household Food Security and the Role of Women.* World Bank Discussion Papers, 96. Washington, DC: World Bank.

Gladwin, C., Thomson, A. M., Peterson, J. S. and Anderson, A. S. (2001). Addressing food security in Africa via multiple livelihood strategies of women farmers. *Food Policy* 26(2): 177–207.

Hallman, K. (2000). Mother–father resources, marriage payments, and girl–boy health in rural. Bangladesh. Unpublished manuscript. Washington, DC: IFPRI.

Hoddinott, J. and Yohannes, Y. (2002). Dietary diversity as a food security indicator. *Food Consumption and Nutrition Division* 136: 1–94.

Jensen, H. and Dolberg, F. (2003). A conceptual framework for using poultry as a tool in poverty alleviation. *Livestock Research for Rural Development* 15(5): 1–17.

Kennedy, E. and Peters, P. E. (1992). *Household Food Security and Child Nutrition: The Interaction of Income and Gender of Household Head.* Development Discussion Paper No. 417. Cambridge, MA: Harvard Institute for International Development, Harvard University.

Kitalyi, A. J. (1998). *Village Chicken Production Systems in Rural Africa: Household Food Security and Gender Issues.* FAO Animal Production and Health Paper No. 142. Rome: FAO. Available at: http://www.fao.org/docrep/003/W8989E/W8989E00.HTM (accessed May 2013).

Kumar, S. K. (1977). *Role of the Household Economy in Determining Child Nutrition at Low Income Levels: A Case Study in Karala.* Cornell University Occasional Paper 95. Ithaca, NY: Cornell University.

Leroy, J. L. and Frongillo, E. A. (2007). Can interventions to promote animal production ameliorate undernutrition? *Journal of Nutrition* 137(10): 2311–2316.

Little, P. D. (1996). *Cooperative Agreement on Human Settlements and Natural Resource Systems Analysis: Cross-Border Cattle Trade and Food Security in the Kenya/Somalia Borderlands.* Binghamton, NY: Institute for Development Anthropology.

Murphy, S. P. and Allen, L. H. (2003). Nutritional importance of animal source foods. *Journal of Nutrition* 133(11): 3932S–3935S.

Nicholson, C. and Thornton, P. K. (1999). The impacts of dairy cattle ownership on the nutritional status of preschool children in coastal Kenya. Paper presented at the Annual Meetings of the American Agricultural Economics Association, Nashville, TN.

Nicholson, C. F., Mwangi, L., Staal, S. J. and Thornton, P. K. (2003). Dairy cow ownership and child nutritional status in Kenya. Paper presented at the Annual Meetings of the American Agricultural Economics Association, Montreal, Quebec, 27–30 July.

Nielsen, H., Roos, N. and Thilsted, S. H. (2003). The impact of semi-scavenging poultry production on the consumption of animal source foods by women and girls in Bangladesh. *Journal of Nutrition* 133(11): 4027S–4030S.

Peacock, C. (2005). Goats – a pathway out of poverty. *Small Ruminant Research* 60(1–2): 179–186.

Peacock, C. (2008). Dairy goat development in East Africa: a replicable model for smallholders? *Small Ruminant Research* 77(2–3): 225–238.

Pinstrup-Anderson, P. (2009). Food security: definition and measurement. *Food Security* 1: 5–7.

Quisumbing, A. R., Brown, L. R., Feldstein, H. S., Haddad, L. and Pena, C. (1995). *Women: The Key to Food Security.* Washington, DC: IFPRI.

Randolph, T. F., Schelling, E., Grace, D., Nicholson, C. F., Leroy, J. L., Cole, D. C. *et al.* (2007). Invited review: Role of livestock in human nutrition and health for poverty reduction in developing countries. *Journal of Animal Science* 85(11): 2788–800.

Reardon, T., Matlon, P. and Delgado, C. (1988). Coping with household-level food insecurity in drought-affected areas of Burkina Faso. *World Development* 16(9): 1065–1074.

Rengam, S. (2001). Women and food security. In Stoll, G. (ed.) *Drawing on Farmers' Experiences in Food Security: Local Successes and Global Failures.* Bonn: German NGO Forum on Environment and Development.

Ruel, M. T. (2003). Operationalizing dietary diversity: a review of measurement issues and research priorities. *Journal of Nutrition* 133: 3911S–3926S.

Savy, M., Martin-Prével, Y., Traissac, P., Eymard-Duvernay, S. and Delpeuch, F. (2006). Dietary diversity scores and nutritional status of women change during the seasonal food shortage in rural Burkina Faso. *Journal of Nutrition* 136(10): 2625–2632.

Scanlan, S. (2004). Women, food security, and development in less industrialized societies: contributions and challenges for the new century. *World Development* 32(11): 1807–1829.

Sinn, R., Ketzis, J. and Chen, T. (1999). The role of woman in the sheep and goat sector. *Small Ruminant Research* 34(3): 259–269.

Sonaiya, E. (2009). Fifteen years of family poultry research and development at Obafemi Awolowo University, Nigeria. In Alders, R. G., Spradbrow, P. B. and Young, M. (eds) *Village Chickens, Poverty Alleviation and the Sustainable Control of Newcastle Disease.* Proceedings of an international conference held in Dar es Salaam, Tanzania, October 2005. Dar Es Salaam: Australian Centre for International Agricultural Research (ACIAR).

Speedy, A. W. (2003). Global production and consumption of animal source foods. *Journal of Nutrition* 133(11): 40–48.

Steyn, N., Nel, J., Nantel, G., Kennedy, G. and Labadarios, D. (2006). Food variety and dietary diversity scores in children: are they good indicators of dietary adequacy? *Public Health Nutrition* 9(5): 644–650.

Swindale, A. and Bilinsky, P. (2006). *Household Dietary Diversity Score (HDDS) for Measurement of Household Food Access: Indicator Guide.* Food and Nutrition Technical Assistance Programme. Washington, DC: AED/USAID.

Uvin, P. (1994). The state of world hunger. *Nutrition Reviews* 52(5): 151–161.

Welch, R. M. and Graham, R. D. (2000). A new paradigm for world agriculture: productive, sustainable, nutritious, healthful food systems. *Food and Nutrition Bulletin* 21(4): 361–366.

WFP (World Food Programme) (2008). *Food Consumption Score.* Interagency Works hop Report, WFP–FAO Measures of Food Consumption – Harmonizing Methodologies, Rome, 9–10 April.

8

MAKING LIVESTOCK RESEARCH AND DEVELOPMENT PROGRAMS AND POLICIES MORE GENDER RESPONSIVE

Jemimah Njuki and Beth Miller

Introduction

Gender mainstreaming has been the primary methodology for integrating a gender approach into research and development effort. Gender mainstreaming is intended to bring the diverse roles and needs of women and men to bear on the development agenda. It is widely recognized that integrating gender perspectives into policies and programs is important to the achievement of all Millennium Development Goals not merely Goal 3 on women's empowerment and gender equality.

Livestock research, development projects and programs, and policies can play a critical role in reducing gender gaps in access to productive resources, income and savings as well as food security if well designed to be gender responsive and to promote women's empowerment. Livestock is a privileged sector for investments to lift poor people out of poverty, and to also promote gender equality, and therefore improve food security. This chapter gives some recommendations on how to make livestock projects, programs and policies more gender responsive. Integrating gender equality goals throughout the livestock value chain will take intentional effort, allocated budgets, and a willingness to move beyond the sureties of the past, but it will give the livestock sector the best opportunity to enhance productivity and food security, and forge collaborations with other sectors to ensure its rightful place in the future of African agriculture.

Rationale for integrating gender in research and development programs

Gender disparities in access to and use of productive resources: As shown elsewhere in this book, there are consistent gender disparities in access to and benefits from technologies, services and inputs across developing countries. Gender-related constraints reflect gender inequalities in access to resources and development opportunities. Although class, poverty, ethnicity and physical location may influence these inequalities, the gender factor tends to make them more severe (Kabeer 2003). Despite the significant roles women play in agriculture and food security they continue to have a poorer command over a range of productive resources and services than men (FAO 2011; World Bank 2001). So, while 40–60 per cent of farmers in sub-Saharan Africa are women, they control less land (women constitute less than 20 per cent of all land holders), and are less likely to use purchased inputs such as fertilizers, improved seeds, mechanical tools and equipment.

Participation in and benefits from markets: Female membership in agricultural marketing cooperatives is generally low, and yet women play a major role in the agriculture sector. A study of membership in dairy cooperatives by the East Africa Dairy Development (EADD) Project found that in Kenya, for all the households that were members of cooperatives, only 27.6 per cent had women registered as members, most of whom were female heads of households. In Rwanda, only 3.1 per cent of households had women registered as members of cooperatives (EADD 2009). Women also lack important information on prices for marketing systems which is often provided by extension agents. Poor female farmers tend to occupy particular niches in the marketing systems. Typically, women are concentrated in small-scale, retail trading, with fewer women involved in trading high up the market hierarchy, for example as wholesalers. Women tend to trade specific commodities such as fresh and highly perishable produce. More generally, agricultural product markets in Africa are gendered because of the gendered access to transport, with the consequence of women traders being concentrated in local markets and men trading in more formal domestic, regional and international markets. Men have better access to information on prices and marketing systems through their intensive marketing networks (Baden 1998).

Men and women are impacted differently by technologies and other interventions: Many agricultural projects still fail to consider the basic questions of differences in the resources, status of men and women, their roles and responsibilities and the potential impacts of interventions on these. Often there is an assumption that as long as there are improved technologies or interventions, they will benefit men and women equally when in fact they may not. Men and women are also impacted differently by and have a role to play in managing emerging threats such as climate change, the HIV/AIDs epidemic, increasing commercialization of resources, and others. Research activities in these themes must take these differential impacts into consideration to ensure that proposed solutions contribute to the current and future improvements in various development outcomes.

A focus on gender can increase the productivity of agriculture and livestock systems, and improve food security and nutrition: Increasing access to productive resources by women to be on a par with those of men would increase farm yields by 20–30 per cent. This in turn would raise agricultural output in developing countries by 2.5–4 per cent, reducing the number of hungry people by 12–17 per cent. Going by the number of hungry people in 2010, such gains in productivity could reduce the number of hungry people by 100–150 million (FAO 2011). Interventions to increase women's access to markets and others that aim to enhance women's income-generating and decision-making ability can lead to improvements in a range of other development outcomes, such as improving child health and nutrition, as well as increasing women's status and eliminating gender differences in asset accumulation. For example, evidence suggests that women spend up to 90 per cent of their incomes on their families while men only spend 30–40 per cent of their incomes on their families (FAO 2011). A large number of studies have linked women's income and greater bargaining power within the family to improved child nutrition status, health outcomes and educational attainment (Smith *et al.* 2003). Findings from the International Food Policy Research Institute's (IFPRI) Gender and Intra-household Research Program have shown the importance of the explicit focus on gender in promoting household poverty reduction. Intra-household dynamics matter as households do not act as one when making decisions. Quisumbing and Maluccio (2000) found that targeting development interventions to more than one person within a household can potentially decrease the effectiveness of development interventions. They show that allocation decisions within a household are not always based on consensus and can undermine women's access to critical resources. Quisumbing (2003) has found that inequality in resource distribution between men and women has both economic and social consequences. This distribution is determined by the "bargaining power" within a household.

Ensuring that both men and women are heard in research and policy processes through meaningful representation in decision-making and policy bodies, in management positions, in research and development is an important component of reducing gender inequalities. Promoting women's organizations and building women's social capital can be an effective tool for women's empowerment. It can be an effective way to improve information exchange and resource distribution, increasing access to resources such as credit, improving women's bargaining power in marketing and managing of their income. Working in groups can help women retain control of income generated from their enterprises. Such organizations can achieve scale as demonstrated by the Self Employed Women's Association (SEWA) in India.

The participation of men and women in agriculture research and development: There is evidence that group diversity leads to better decision outcomes, better performance, creativity and innovation, and this has been shown in a variety of settings, occupations and organizations (Hamilton *et al.* 2004; Pelled *et al.* 1999). Diversity is beneficial because a variety of opinions, backgrounds and thinking styles, and their integration into the solution, are what contributes to better decision outcomes. From a gender perspective, research has found a correlation between the presence of women in higher management and performance of the organization; and having gender diversity in teams has been found to double performance (Mannix and Neale 2005). Women, however, face different constraints in the workplace that limit them from moving into decision-making positions. Organizational practices and prejudices, including hiring and incentive systems, can often work against women. A survey on female participation in African agricultural research and higher education done in 2007/8 found that women are still under-represented in (agricultural) science and technology (S&T) systems in most countries. The study found that the female share of the research workforce was about 23 per cent, with only 14 per cent being in management positions. Women are less represented in the high-level research management and decision-making positions compared with their male counterparts. Women's participation declines as they progress along the career path (Beintema and Di Marcantonio 2010).

Key steps for making livestock programs and policies more gender responsible

Place livestock and gender in a wider livelihoods context

Local culture and attitudes, as well as the political and natural environment, affect the decision-making options and incentives for livestock keepers, yet professional training, policy documents and field activities rarely reflect this. Livestock officials, veterinarians, economists, researchers and animal scientists must understand the social context of the value chain, and why and how to intentionally include women and other marginalized groups in training and information exchanges, market participation and policy development (Rushton 2009).

Institutions are social structures governing the behaviour of a set of individuals within a given human community, with rules and enforcement. Institutions range from formal governments to businesses, NGOs, the family and the market, and each will have its own culture or "internal rules". Women and men are embedded in a system of institutions that define rules of action and create incentives and punishments. Such systems differ by location and over time so that incentives vary. Although individuals differ in their preferences and capacities, all individual choices are contextualized by their society and its institutions. Therefore, development and policy models which assume autonomous individuals maximizing personal benefit but which do not consider the social context can come to misleading conclusions and produce flawed policies. This is especially true for women, because most African cultures impose stricter control over women's behaviour and choices compared to men of the same age, class and ethnicity.

Livestock constitute only one of the economic and non-economic activities that households engage in, and must therefore be looked at in that context. Livestock interventions will affect not only livestock-related activities but other activities as well. For example, livestock interventions that increase the time women spend on livestock keeping will have implications for their time and the care they can give to children; they will affect their nutrition-related activities as well as their leisure time. This interrelationship means that livestock interventions and policies have to be looked at from a broader perspective, taking into account linkages with other sectors.

Using gender transformative approaches in agriculture and livestock development

Gender transformative approaches help guide in achieving both sectoral outcomes (e.g. increased agricultural production, improved food security) and gender equality outcomes. By working on both sectoral and gender equality outcomes, livestock programs have a better chance of achieving sustainable change.

Gender transformative approaches require that, in addition to integrating gender in the programmatic approaches, gender and power inequalities are also addressed. Figure 8.1 shows a gender integration continuum. Those working in livestock programs should strive towards gender transformative programming and should not be implementing "gender exploitative" programming. *Gender blind* refers to the absence of any proactive consideration of the larger gender environment and specific gender roles affecting program/policy beneficiaries. Gender blind programs/ policies would give no prior consideration to how gender norms and unequal power relations affect the achievement of objectives, or how objectives impact on gender. *Gender aware* refers to explicit recognition of local gender differences, norms and relations, and their importance to outcomes (could be health-, education-, livelihoods-related outcomes) in project design, implementation and evaluation. This recognition derives from the analysis or assessment of gender differences, norms and relations in order to address gender equity in outcomes.

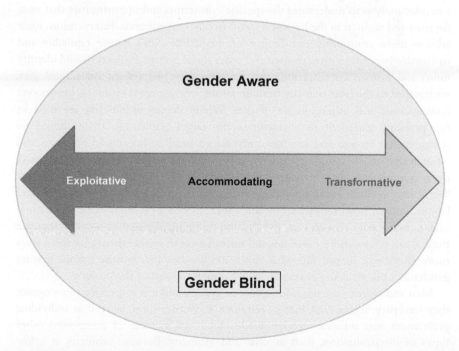

FIGURE 8.1   Gender and women's empowerment continuum

Gender exploitative refers to approaches to project design, implementation and evaluation that take advantage of rigid gender norms and existing imbalances in power to achieve the program objectives. Livestock programs with nutrition components, for example, may focus on working with women-only groups to disseminate the importance of milk to children's nutrition based on women's roles in household nutrition. *Gender accommodating* approaches, on the other hand, acknowledge the role of gender norms and inequities, and seek to develop actions that adjust to and often compensate for them. While such projects do not actively seek to change the norms and inequities, they strive to limit any harmful impact on gender relations. After a gender analysis, program staff realize that men tend to raise cows but women raise chickens. Cows are not traditional for women to raise, and no women own cows. The program recognizes these gender differences and implements a separate strategy for men and women, with a men's one focused on cows and a women's one focused on chickens. *Gender transformative* approaches actively strive to examine, question and change rigid gender approaches, norms and imbalances of power as a means of reaching sectoral outcomes (e.g. increasing agricultural productivity, livelihoods, food security, market engagement, nutrition) while also promoting more gender equitable objectives.

Conduct a gender analysis to inform program design

Livestock program and policy designers and implementers should always start with a gender analysis to understand the specific constraints and opportunities that exist for men and women in the livestock sector in different contexts. Interventions must address these constraints, reduce gender inequalities, and ensure equitable and sustainable benefits to men, women and other social groups. Projects should identify men's and women's needs, constraints, opportunities, preferences for technologies, with regard to the issue of focus, from literature review, expert opinions, pre-project consultations and other sources of data. Where demographic data are used in the problem statement to characterize the target population these should be disaggregated by age and sex (not only sex of head of household but men and women farmers). In analysing the context in which the project will be implemented, the gender relations and inequalities that exist should be identified and documented. These may include constraints in access to resources and assets, information and labour. Women, however, play important roles in livestock production, environmental management and other sectors, and this should be highlighted in order to increase their access to resources, capacities and information to enable them play these roles more effectively. Identifying what the issues are is a prerequisite for integrating gender in a practical and systematic way through the rest of the project.

Men and women are not homogeneous groups, and it is important to recognize their diversity across class, ethnic, religious and other lines, as well as individual preferences and abilities. Gender disadvantage can increase or compound other types of discrimination, such as class and ethnicity. Because ethnicity is a key defining characteristic in Africa, it drives discrimination as well as conflict, state formation, political alliances and economic choices (Hickey and Du Toit 2007). For poor women belonging to ethnic minorities, such as Maasai women in Kenya, claiming productive resources is doubly challenging, as is exercising political power to promote favourable policies. Roles and responsibilities can change with age. In Fulani areas in Nigeria, a young girl making butter under her mother's supervision eventually becomes the manager of her own small-scale dairy operation and, with increasing age, may also take on increasing responsibilities (Waters-Bayer and Letty 2010). The gender analysis should strive to analyse and understand differences across men and women, and among men and women, as well as across generations.

Several frameworks and tools exist for carrying out gender analysis. The FAO Socio-economic and Gender Analysis (SEAGA) is an approach to development based on an analysis of socio-economic patterns and participatory identification of women's and men's priorities. The objective of the SEAGA approach is to close the gaps between what people need and what development delivers. The approach combines socio-economic analysis and gender analysis to enable learning about community dynamics, including the linkages among social, economic and environmental patterns. It clarifies the division of labour within a community, including divisions by gender and other social characteristics, and it facilitates understanding of resource use and control, and participation in community institutions. SEAGA focuses on three domains of analysis: the development context, livelihood analysis and stakeholder analysis. Common gender analytical frameworks including the Harvard Analytical Framework, also known as the Gender Roles Framework (Overholt et al. 1985), the Moser Gender Planning Framework (Moser 1993), the Gender Analysis Matrix (Parker 1993), the Women's Empowerment Framework (WEF) and the Social Relations Approach (Kabeer 1994) provide good starting points for conducting a gender analysis. These tools need to be adapted to the local context in which they are used.

The CARE (Cooperative for Assistance and Relief Everywhere) good practices framework on gender analysis (CARE 2012) outlines three key phases of gender analysis to explore gender dynamics from broader to local contexts: preliminary foundations to understand the context in which to situate the gender analysis; core enquiries of gender which cut across CARE's women's empowerment domains of agency, structure and relations; and applying gender analysis to programming focusing on key immediate rights that affect women's conditions (practical rights) as well as the needed transformation in structures and relations to pursue gender equality (strategic interests).

More recent tools for gender analysis include gender and value chain analysis tools. The gender-sensitive value chain analysis tools by the International Labour Organization (ILO; Mayoux and Mackie 2009) provides tools and methods, originally for incorporating gender concerns into the different stages of value chain analysis and strengthening the links essential for gender equality and promoting sustainable pro-poor growth and development strategies. The *Integrating Gender in Agricultural Value Chains* handbook (Rubin et al. 2009) provides a phased process for integrating gender into agricultural value chains. The handbook provides a five-phase approach for analysing and integrating gender into value chain analysis and development: mapping gender roles and relations along the value chain; moving from gender inequalities to gender-based constraints; assessing the consequences of gender-based constraints; taking actions to remove gender-based constraints; and measuring the success of actions.

Integrate gender in project and policy design and implementation

Integration of gender into projects, programs and activities should use the project cycle to ensure that gender is integrated in all key aspects of the project. Gender aspects should be an integral part of the problem analysis, project goals and objectives. It should be systematically and practically included in the operational plan by translating it into concrete activities and relevant indicators. There is evidence that such systematic integration of gender concerns in projects leads to better outcomes. Kristjanson *et al.* (2010) found that interventions (or policies) with positive impact on women were those that focused on women from the beginning, rather than simply adding women into existing activities. Projects already designed around men's priorities are often inappropriate for women, so investment in institutional capacity for redesign, along with new activities, is often necessary.

The project cycle is an appropriate tool for integrating gender as it ensures that the problem analysis is thorough and done from a gender perspective; stakeholders are clearly identified and analysed, including their gender capacities; objectives are relevant, gender responsive and clearly stated; outputs and objectives are logical and measurable; men's and women's strengths and weaknesses have been identified; assumptions are taken into account; monitoring concentrates on verifiable targets and outputs; evaluations identify "lessons learnt" and integrate them into the cycle for similar succeeding projects; and sustainability is defined, not essentially by "organizational continuity", but primarily by the continuous "flow of benefits". Integrating gender in this systematic process ensures gender is integrated at every stage.

Several guidelines and checklists exist for guiding programs on how to integrate gender within the program cycle. At the International Livestock Research Institute (ILRI), the strategy for integrating gender takes a very livestock-oriented focus, looking at key gender issues in livestock and how the institute and other research and development organizations can better address the needs of both men and women in their work (ILRI 2012).

Very often for projects and programs, the question has always been "Why is gender important?" which gives a supply-driven impetus for integrating gender, especially from gender experts within organizations. Given the evidence that exists, the burden of proof needs to shift and projects have to demonstrate why gender is not important if they do not want to integrate gender systematically.

A gender blind priority-setting process is not likely to yield a gender-balanced project portfolio. Addressing gender issues in priority setting requires examining which crops and animals, and which markets are selected for research, and what women's roles in and potential benefits from these are. If priority-setting processes are done with stakeholders, both men and women should be involved in the process. Once priorities are set, projects should define gender-responsive goals and objectives. This can be done at two levels: (i) gender as a stand-alone research objective/ research topic (i.e. strategic gender research); or (ii) gender as a cross-cutting thematic research area, in which gender analysis is used to inform and deepen other research themes. Some stand-alone gender objectives include objectives such as: reducing gender disparities in access to resources, livestock assets or markets; increasing collective action by women, etc. If gender is a cross-cutting area across other research objectives, this should be clearly stated in the objectives or research questions. Gender blind objectives lead to gender blind activities and implementation approaches. Making objectives or research questions gender responsive goes beyond adding such statements as "including women, or especially women" at the end of the objective. A gender-responsive objective could be: increase incomes of men and women from livestock; improve the nutrition of men, women, boys and girls within smallholder households; develop technologies that address men and women's constraints, among others.

Research and development implementation approaches should address key gender issues identified in the gender analysis. This will involve integration of gender strategies to address existing gender inequalities and build on opportunities for men, women, boys and girls and other community members. This may involve strategies targeted at men and women, or targeted at women only. Targeting of women is sometimes necessary to address existing gender imbalances. The interventions could be broadly categorized into two kinds: immediate and practical interventions, and strategic interventions. Immediate and practical interventions are those that address immediate concerns and often have short-term results. Examples of these could include organizing women to access markets, building capacity, increasing access to basic resources such as water, health, etc. Strategic interventions are often of a longer-term nature and aimed at changing structural gender relations that cause inequalities. These could, for example, be changing legal provisions for land ownership so that women can own land.

Involvement of men and women scientists, and development staff in the implementation of the project and in decision-making processes is critical. Programs and projects need to go beyond getting men and women to participate in project activities to ensuring that they benefit and that there is a transformation of unequal gender relations to achieve equity.

Program plans should describe all the activities that will be carried out to deliver on the gender objectives and the gender strategy. It is not enough to have gender-responsive goals and objectives if these are not followed by activities to achieve them. In developing the work plan, project teams should ask themselves whether the gender-specific activities are sufficient to deliver on the goals and objectives. If the objective is to increase women's access and adoption of a technology, what are the gender-specific activities that will make this happen? If it is to increase men and women's income, what are the specific activities that will lead to women's management of income? Each project/program should have appropriate staffing levels, including expertise to implement the gender activities and strategy. This can be new expertise or drawn from the project staff or partners. Gender training for all staff to create awareness and build basic skills for facilitation and for integrating gender is critical.

The specific costs allocated to gender activities for staffing, implementation of gender activities and capacity building should be clearly specified and allocated. This ensures that gender is not an add-on activity for which no budget is allocated. In research for a development project a minimum of 5 per cent of the budget should be dedicated to gender research, activities and capacity building.

TABLE 8.1 Tools for integrating gender within the project cycle

Stages of the project cycle	Potential tools for integrating gender
Problem and context analysis	Secondary data
	Existing national, regional, local data sets and studies, e.g. national agriculture surveys
	Qualitative approaches that integrate gender analysis frameworks and tools
	Gender and value chain analysis tools
	Gender and risk assessment tools
	Rapid/qualitative appraisals
Setting priorities, identifying goals and objectives	Using secondary data
	Using existing knowledge of key gender issues related to the context of the research
Research and development	Some examples may include:
	Participatory technology/value chain development with men and women farmers
	Group-based approaches, e.g. Village Savings and Loan Associations with men and women groups
	Gender training for staff and communities
	Engaging men and boys for equitable gender relations
Work plan development	Three sets of activities should be distinguished in the work plan
	Activities where gender is integrated or mainstreamed, e.g. gendered value chain analysis
	Activities where gender is a strategic approach or research area
	Activities directed at certain groups of people, e.g. women, female-headed households, youth, e.g. formation of women groups to increase women's participation in marketing activities
Budgeting	Gender-responsive budgeting
	Activity-based budgeting where gender activities, e.g. analysis, are budgeted for
Monitoring and evaluation	Having gender-specific outputs, outcomes and impacts
	Disaggregating all indicators and data collection by gender
	Measuring women's empowerment

Gender in monitoring, evaluation and impact assessment

Programs should articulate and present a plan for a gender-responsive monitoring and evaluation (M&E) system for strategy-level goals, outcomes, outputs and activities, as well as thematic research areas, and articulate clear plans on how the results of gender-responsive M&E will be systematically used for learning about what works in gender and agriculture, assessing progress towards achieving gender equality and for informing programs and policies. The differential impacts of programs on women and men can only be identified if M&E mechanisms are gender responsive and measure changes in men's and women's situation.

In order to measure how well a development project or program has scored in its gender targets and if its results relating to gender equality have been achieved, monitoring, evaluation and impact assessment systems must be gender sensitive. This will allow for measuring gender-related changes in society over time. They can also make visible often hard-to-find or assumed issues, such as men's and women's roles in productive and reproductive activities, data which can be useful for national planning. Information produced from such systems and processes can be used to advocate for gender equality and advance the agendas of women's empowerment.

Gender equity and women's empowerment as a goal or outcome

Gender equality and women's empowerment can be the goal or an outcome of development programs and policies. There have been attempts to develop indicators to measure empowerment of women or other groups. However, empowerment can have many meanings, is complex, and often includes people's subjective feelings of power or lack of agency. It is thus important to be clear about exactly what empowerment indicators measure and show, and to complement this with qualitative gender analysis.

The Bangladesh Rural Advancement Committee (BRAC) and Grameen Bank have used eight indicators to measure women's empowerment: mobility, economic security, ability to make larger purchases, involvement in major household decisions, relative freedom from domination within the family, political and legal awareness, and involvement in political campaigning and protests (Oxaal 1997).

The Women's Empowerment in Agriculture Index (WEAI), developed for USAID (US Agency for International Development) by IFPRI and Oxford University, seeks to capture women's empowerment and inclusion levels in the agricultural sector, to raise the status of women in agriculture, improve nutrition and decrease poverty. The index considers five factors to be indicative of women's overall empowerment in the agricultural sector:

Decisions over agricultural production
Power over productive resources such as land and livestock
Decisions over income
Leadership in the community
Time use

Women are considered empowered if they score adequately in at least four of the components (IFPRI 2012). The index uses individually based data of men and women in the same households to calculate both a women's empowerment index and a gender parity index.

Gender-sensitive indicators and collecting sex-disaggregated data

Sex-disaggregated statistics give the straightforward numbers of males and females in a given population, while gender data can reveal the relationships between women and men that underlie the numbers. Gender-sensitive indicators provide evidence of (changes in) the situation and position of women, relative to the status of men.

There needs to be an understanding among program teams and policy makers that sex disaggregation does not have to place women and men in opposition to one another, and cannot assume that they are collections of isolated, atomized individuals with only individual and separate interests. Data collection must also place them within their wider social contexts of gender, age, class and other identities that influence their relations with others (Okali 2011). The sex disaggregation should go beyond the common assumption that collecting data from male- and female-headed households constitutes sex disaggregation and gender analysis. It has to reflect disaggregation within households – men and women, boys and girls – and an analysis that shows the relationships between and among them. Collection of sex-disaggregated data is not enough if this data is not analysed and used to inform policies and programs. Evidence-based policy is crucial to ensure that the livestock sector plays a significant role in economic growth, food and nutrition security, and reduction of gender inequality. Baseline data on gender relations or the gender situation should be carried out before the project or program is implemented to provide a basis for assessing the results and impact of a program and policy.

Who is asked for information and who asks for the information is an important consideration. The collection and analysis of the information gathered is not a gender neutral process and is subject to gender bias and gender-laden cultural attitudes. Sometimes it might be more appropriate to have women interviewers, or interviewers of the same sex as interviewees. Interviewers might be less comfortable talking with one sex or another, especially in some cultures. Teams collecting M&E and impact data should be gender aware and should be trained in gender. Data collectors who are not gender aware may disregard certain important data or play down the importance of particular gender differences. Even when data has been disaggregated during the collection, these differences may not be retained during the analysis, interpretation and reporting of the data if those carrying out the task have not been trained on gender.

Indicators should be developed at the different levels of the results chain. They should include input indicators (inputs), process indicators (activities and how these lead to outputs) and progress or outcome indicators (outcomes and impacts).

TABLE 8.2 Examples of gender indicators for livestock projects

Activities	Outputs	Outcomes	Impacts
Men's and women's level of participation in project activities	Men and women's preferences for technologies/services	Adoption rates/use of technologies and services by men and women	Changes in income and equitable share of income among men and women
Implementation of specific initiatives to address gender issues in access to resources, information, assets, capacity in livestock sector	Number of staff trained on gender	Market participation by men and women	Contribution of livestock to women's/men's income and subsistence
	Number of extension messages produced/ disseminated on gender issues		
Funds allocated/ disbursed for capacity building training on gender for research and extension staff	Number of women and service providers		Changes in nutritional status and availability of milk and animal protein by men and women
	Number of women participating in and benefiting from producers' associations and cooperatives		
Women's level of participation in producers' cooperatives; women's group for collection and marketing	Changes in marketing network and patterns		Changes in gender asset disparity

Within a program or organization, the process for developing these indicators should be participatory, involving all staff in the program, so that there is an agreement that these are important for the program or organization. This process should include how the data or information on these indicators will be used within the program or organization. For example, at ILRI, a core set of six gender-sensitive outcome and impact indicators were agreed on for measuring the impacts of ILRI projects and programs (Njuki *et al*. 2011). Some examples of these are shown in Table 8.3.

For each of these impact and outcome areas and the indicators, formats for data collection, and the calculation and presentation of the indicators is included.

Involving men and women in gender-responsive M&E and impact assessment

A participatory M&E process is one in which the target groups have genuine input into developing indicators to monitor and measure change. If successful, this allows for the M&E process to be "owned" by the group rather than imposed on them by outsiders. Often, men and women have different priorities and different indicators for measuring change. Involving both men and women ensures these differences are captured and taken into account in designing M&E systems, as well as in the implementation of the program, project or policy. In response to a question of how

TABLE 8.3 Examples of core gender, livestock and livelihood indicators for livestock projects (developed by ILRI)

Outcome and impact area	Indicators
Asset accumulation	*Domestic assets* • Household domestic asset index • % of women who own different assets • Gender asset disparity *Livestock* • % of households in where women own livestock (by and across species) • % of livestock in survey owned by women (not using Tropical Livestock Units [TLUs]) • % of total TLUs under women's ownership (by and across species) • Average number of livestock owned by women per household (by and across species)
Income	• Annual farm and off-farm income • % of total annual income managed by women (total and by source) • Cash income from livestock and livestock products • Contribution of livestock to total farm/household income • % of livestock income managed by women (total and by source)
Food security	• Individual Dietary Diversity Score for female adult, male adult, female child under 5 and male child under 5 • Proportion of men, women, girls and boys consuming at least one animal source food per day • Number of months of adequate household food provisioning in male- and female-headed households
Labour use in livestock systems	• Amount of labour used in livestock, by activity and gender
Access to inputs, services and technologies	• Percentage of households with access to a technology or input • Percentage of households who have used, in past 12 months, a technology or input • % of women with access to different technologies or inputs • Women's decision-making on use of technology or inputs (% of households where women made the decision to use a specific technology or input) • % of households with savings in formal and informal savings mechanisms • % of women with savings in formal and informal savings mechanisms • % of households who have taken a loan in the last five years

TABLE 8.4   Indicators prioritized by men and women in Malawi for increased incomes

Men	Women
• New income-generating activities initiated	• Women having bank accounts in their own names
• Men not drinking traditional beer	• Children going to secondary school
• Men marrying a second wife	• Good food (breakfast, good quality tea)
• Iron sheet roofed houses	• Women going to market weekly
	• Better clothing – women wearing new *khangas* (wraps), *kodokodo* (pointed shoes)
	• More women participating in meetings as they will have a change of clothes

the program implementers would know whether men and women had increased their incomes, they gave different indicators as shown in Table 8.4.

Using multiple tools and methods to measure change

Gender issues are inextricably linked to cultural values, social attitudes and perceptions. This means that measuring them will require multiple methods, qualitative and quantitative. While quantitative approaches give an indication of how much change has happened, qualitative methods are especially useful to understand social processes, why and how a particular situation measured by indicators has taken place, and how such a situation could change in the future. These methods provide an in-depth understanding of what is changing, why and for whom, and with what implications. Tools such as participatory impact diagrams can provide an opportunity for men and women to discuss the impacts of interventions on their lives, both positive and negative (see Figure 8.2).

Other useful tools include focus group discussions with men and women, ranking and scoring tools, and ethnographic techniques among others. A careful selection of a combination of tools can yield important lessons on how change is happening to men and women, boys and girls and other community members.

Conclusions

Livestock provides an opportunity for women's economic empowerment and for reducing gender disparities in ownership of assets and resources. For livestock programs to economically and socially empower women, they need to combine both sectoral objectives (e.g. increasing productivity, improving livelihoods, food security, market engagement, nutrition) and the promotion of more gender-equitable objectives, such as women's empowerment or equitable agriculture systems.

The project cycle is a systematic representation of the process of formulating an intervention from inception to conclusion. The stages of the project cycle provide a structure that ensures that the problem analysis is thorough; stakeholders

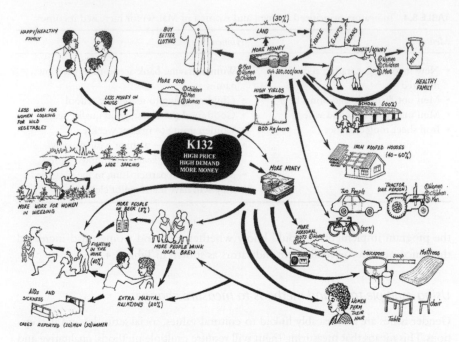

FIGURE 8.2 Impact of improved bean varieties by a mixed group of farmers in Nabongo parish, Uganda

Source: David (1999)

are clearly identified and monitored; quality assurance is built in; objectives are relevant to problems and clearly stated; outputs and objectives are logical and measurable; beneficiaries' strengths and weaknesses have been identified; assumptions are taken into account; monitoring concentrates on verifiable targets and outputs; evaluations identify "lessons learnt" and integrate them into the cycle for similar succeeding projects; and sustainability is defined, not essentially by "organizational continuity", but primarily by the continuous "flow of benefits" to improve local livelihoods.

Integrating gender in programs in a systematic way through the project cycle ensures that women's and men's needs and priorities are addressed, their constraints are addressed and any interventions have positive impacts on both men and women. Using gender transformative approaches leads to addressing unequal power relations between men and women. Capacity building, not only in understanding gender issues and conducting gender analysis, but also in behaviour change and facilitation skills, is a prerequisite for the use of gender transformative approaches. Working with men and women to change unequal gender relations and define change from their own perspectives, and engaging men and boys to support women's empowerment, can lead to the desired long-term changes in gender inequality.

References

Baden, S. (1998). Gender issues in agricultural liberalization. Topic paper prepared for Directorate General for Development (DGVIII) of the European Commission. Bridge Report No. 41. Brighton: Institute for Development Studies, Sussex University.

Beintema, N. M. and di Marcantonio, F. (2010). *Female Participation in African Agricultural Research and Higher Education: New Insights.* IFPRI Discussion Paper 00957. Washington, DC: IFPRI.

CARE (2012). Good practices framework on gender analysis. Available at: http://www.care.org (accessed May 2013).

David S. (1999). *Beans in the farming system and domestic economy of Uganda: a tale of two parishes.* Kampala: CIAT.

EADD (2009). *Gender, Dairy Production and Marketing.* East Africa Dairy Development Project. Nairobi: ILRI.

FAO (2011). *State of Food and Agriculture – Women in Agriculture: Closing the Gender Gap.* Rome: FAO.

Hamilton, B., Nickerson, J. and Owan, H. (2004). Diversity and productivity in production teams. SSRN 547963. Available at: http://papers.ssrn.com/sol3/papers.cfm?abstract_id=547963 (accessed May 2013).

Hickey, S. and Du Toit, A. (2007). *Adverse Incorporation, Social Exclusion and Chronic Poverty.* CPRC Working Paper No. 81. Available at: http://www.chronicpoverty.org/publications/details/adverse-incorporation-social-exclusion-and-chronic-poverty (accessed May 2013).

IFPRI (2012). *Women's Empowerment in Agriculture Index* (WEAI). Washington, DC: IFPRI.

ILRI (2012). *Strategy and Action Plan for Mainstreaming Gender at ILRI.* Nairobi: ILRI.

Kabeer, N. (1994). *Reversed Realities: Gender Hierarchies in Development Thought.* London: Verso.

Kabeer, N. (2003). *Gender Mainstreaming in Poverty Eradication and the Millennium Development Goals.* London: IDRC/Commonwealth Secretariat. Available at: http://www.thecommonwealth.org/shared_asp_files/uploadedfiles/%7Beeea4f53-90df-4498-9c58-73f273f1e5ee%7D_povertyeradication.pdf (accessed May 2013).

Kristjanson, P. *et al.* (2010). *Livestock and Women's Livelihoods: A Review of the Recent Evidence.* Discussion Paper No. 20. Nairobi: ILRI. Available at: http://mahider.ilri.org/bitstream/10568/3017/2/Discussion_Paper20.pdf (accessed May 2013).

Mannix, E. and Neale, M. A. (2005). What differences make a difference? The promise and reality of diverse teams in organizations. *Psychological Science in the Public Interest* 6(2): 31–55.

Mayoux, L. C. and Mackie, G. (2009). *Making the Strongest Links: A Practical Guide to Mainstreaming Gender Analysis in Value Chain Development.* Addis Ababa: ILO.

Moser, C. (1993). *Gender Planning and Development: Theory, Practice and Training.* London: Routledge.

Njuki, J. (2011). Gender and livestock value chains in Kenya, Tanzania and Mozambique. Paper presented at the Workshop on Gender and Market Oriented Agriculture, Addis Ababa, 31 January–2 February.

Okali, C. (2011). *Integrating Social Difference, Gender and Social Analysis into Agricultural Development.* Brighton: Institute for Development Studies, Sussex University.

Overholt, C., Anderson, M. B., Cloud, K. and Austin, J. (1985). *Gender Roles in Development Projects: A Case Book.* West Hartford, CT: Kumarian Press.

Oxaal, Z. (1997). *Gender and Empowerment: Definitions, Approaches and Implications for Policy.* BRIDGE Report No. 40. Brighton: BRIDGE, IDS.

Parker, A. R. (1993). *Another Point of View: A Manual on Gender Analysis Training for Grassroots Workers*. New York: UNIFEM.

Pelled, L. H., Eisenhardt, K. M. and Xin, K. R. (1999). Exploring the black box: an analysis of work group diversity, conflict and performance. *Administrative Science Quarterly* 44(1): 1–28.

Quisumbing, A. R. (2003). *Household Decisions, Gender, and Development: A Synthesis of Recent Research*. Washington, DC: IFPRI.

Quisumbing, A. and Maluccio, J. (2000). *Intra-household Allocation and Gender Relations*. Washington, DC: IFPRI.

Rubin, D., Manfre, C. and Nichols Barrett, K. (2009). *Integrating Gender in Agricultural Value Chains (Ingia-VC) in Tanzania*. Greater Access to Trade (GATE) project. Washington, DC: USAID.

Rushton, J. (2009). *The Economics of Animal Health and Production*. Oxfordshire: CABI.

Smith, L., Ramakrishnan, U., Ndiaye, A., Haddad, L. J. and Martorell, R. (2003). *The Importance of Women's Status for Child Nutrition in Developing Countries*. Research Report 131. Washington, DC: IFPRI. Available at: http://www.ifpri.org/publication/importance-womens-status-child-nutrition-developing-countriess (accessed May 2013).

Waters-Bayer, A. and Letty, B. (2010). Promoting gender equality and empowering women through livestock. In Swanepoel, F. J. C., Stroebel, A. and Moyo, S. (eds) *The Role of Livestock in Developing Communities: Enhancing Multifunctionality*. Wageningen, The Netherlands: CTA.

World Bank (2001). *Engendering Development: Through Gender Equality in Rights, Resources, and Voice*. Washington, DC: World Bank.

9

CONCLUSION

Improving the design and delivery of
gender outcomes in livestock research
for development in Africa

Pascal Sanginga, Jemimah Njuki and Elizabeth Waithanji

A new momentum on gender equality in agricultural research and development

There is no debate about the importance of closing gender gaps in agriculture and food security. The research upon which this book is based was conducted at a time when the incentives and initiatives for gender integration in agriculture research and development were stronger than ever because of the rising global consensus and public action by international organizations, national government and non-governmental organizations (NGOs) on the importance of women's empowerment in economic development and poverty alleviation. Three recent high-profile publications – *The Gender in Agriculture Sourcebook* (World Bank, FAO and IFAD 2009), *The State of Food and Agriculture 2010–11: Women in Agriculture – Closing the Gender Gap for Development* of the United Nations Food and Agriculture Organization (FAO 2011) and the World Bank's *World Development Report 2012: Gender Equality and Development* (World Bank 2012) – summed up the evidence showing that addressing gender inequalities and empowering women are vital to meeting the challenges of improving food and nutrition security, and enabling poor rural people to overcome poverty. In addition, the first ever African Human Development Report, *Towards a Food Secure Future* (UNDP 2012), concludes that building a food-secure future for Africans will require focus and actions in the critical areas of empowering women and the rural poor in order to increase the productivity of smallholder farmers, advance nutrition among children, and build resilient communities and sustainable food systems (UNDP 2012). Several other international organizations have revamped and published their gender equity strategies over recent years.

Several publications have documented considerable progress in sub-Saharan Africa in narrowing of many gender gaps and empowering women in several sectors, namely: education, health, labour market opportunities and political representation. Although the field of gender in agriculture has a strong scientific basis and there is ample field experience, progress on closing gender gaps in agriculture and food security has not kept pace with other sectors. Where pilot projects have documented success in reducing gender disparities, the outcomes have not been sufficiently sustained and widespread (CGIAR 2012). Gender inequalities persist for rural women in the agricultural sector, where women play significant roles but where they continue to face significant constraints and barriers. A growing and rich scholarship in gender and agriculture, spanning more than four decades, has accumulated empirical evidence on the diversity of women's and men's dynamic roles and the responsibilities they take for improving the four dimensions of food security: availability, access, utilization and stability; as well as the challenges, inequalities and opportunities that women particularly face in improving their livelihoods and benefiting from development interventions (for a review see Meinzen-Dick *et al.* 2010). While there is a growing number of excellent studies on gender in crop management and crop value chains, the dearth of information and data on gender in livestock is particularly notable. Livestock are one of the most important assets for women, yet little evidence exists on the extent, nature and processes of ownership of livestock by women, and their decision-making power over livestock.

It is well recognized that advances in agriculture and food security can only come about with explicit gender focus owing to the feminization of agriculture (Chant 2010; Meinzen-Dick *et al.* 2010; Quisumbing and Meinzen-Dick 2001) and the need for social equity. It is also recognized that in order to have the greatest impact, agricultural research and development programs must target and benefit small-scale women farmers who represent the majority of rural poor populations in developing countries.

The studies reported in this book sought answers to important but rarely addressed questions, namely: What do we know about the gender-differentiated preferences and ownership of different livestock species? What are the gendered patterns in livestock decision-making and income management? What are the food security implications of gendered control of livestock and livestock income? This book combines the latest knowledge on gendered livestock asset gaps and decision-making about these assets. The studies are based on the findings of empirical analyses of sex-disaggregated data to establish the extent of women's and men's ownership and control of livestock and livestock products, their preferences, participation in and benefits from livestock value chains, income management and allocation, and decision-making on livestock, their products and income generated from their sales as well as their impacts on food and income security and empowerment of women. The aim of this concluding chapter is to distil the main findings and lessons from the previous chapters and to make some general reflections about how to move gender in livestock research and development to the next frontiers.

What livestock matter for women?

Livestock play an important role in feeding billions of people, sustaining millions of smallholder food producers, providing income sources, intensifying small-scale mixed farm production and making productive use of dryland resources, reducing vulnerabilities of pastoral systems, providing a buffer against periodic hunger and drought, and influencing climate change (for good or bad) as smallholder livestock keepers can make their livestock production more efficient and profitable (FAO 2011; ILRI 2012a). In rural Africa, livestock is one of the most valuable agricultural assets. It represents a primary source of income and wealth accumulation, assurance and investments that are more important than business and housing for millions of poor people. Results reported in this book confirm that livestock contribute significantly to the asset base of the poor, contributing up to 85 per cent of the total movable assets in Mozambique, and more than 50 per cent in Kenya and Tanzania. It is recognized that ownership of livestock assets is an important aspect of women's economic empowerment because this increases their participation in household decision-making and the extent to which women can respond to and benefit from marketing opportunities and incentives.

Several studies (Delgado 2003; Delgado *et al.* 1999; ILRI 2012a; Pica-Ciamarra and Otte 2009) have termed the increasingly high demand for animal source foods and other livestock products due to urbanization, demographic and social changes, the "livestock revolution". This "livestock revolution" provides economic opportunities that will benefit more than half a billion smallholders who depend on livestock. However, as in the "Green Revolution", these studies do not highlight the gender implications of the livestock revolution and neglect the specific needs of women as livestock keepers, owners, decision-makers, processors and value chain actors. Yet it is estimated that two-thirds of the 1 billion poor people who depend on livestock for their livelihoods are rural women (Staal *et al.* 2009). While livestock could increase poor people's ability to move out of poverty, the women among them face gender-based biases and challenges in livestock value chains. Within the gender and livestock literature, there is a distinct pattern of gender differentiation in ownership and preference of animals according to their type. A general pattern around the world is that women tend to own more poultry, and have more control and decision-making power over poultry and other small animals (FAO 2011). It is argued that women do not own, control and benefit from large livestock (Deere *et al.* 2012; FAO 2011; Kristjanson *et al.* 2010; Mupawaenda *et al.* 2009; Wooten 2003), which are essentially men's domain. Empirical results from several chapters in this book nuance these assumptions, which are routinely presented in livestock and gender studies. While there is evidence that the majority of rural women in Kenya, Tanzania and Mozambique own poultry, a considerable proportion of women also own large livestock, reaching 40 per cent in Mozambique. Results further showed that small ruminants (sheep and goats) and pigs contributed negligibly to women's Tropical Livestock Units. On the contrary, cattle contributed more to women's Tropical Livestock Units than goats, sheep and chickens. Men and women preferred dairy cattle and dairy goats equally, but women consistently owned fewer dairy cattle and goats than men, sometimes up to six times fewer. These findings are in line with Quisumbing *et al.*'s (2001) and Dolan's (2001) findings that the gendered patterns of livestock ownership and preference are also changing, and gender ownership and management of livestock may be less rigid than they initially appear.

Beyond female-headed households

Studies that have analysed the gender dimensions of livestock ownership have often been conducted at the household level and are often limited to distinguishing male- and female-headed households (Kristjanson *et al.* 2010; Mupawaenda *et al.* 2009; Saghir *et al.* 2012). Recent literature demonstrates that gender analysis that only distinguishes male- and female-headed household heads is unsatisfactory and incomplete since "it reduces gender to the sex of the household head and does not allow for analysis of the relative position of men and women within households where adults of both sexes are present" (Deere *et al.* 2012: 506). The reliance on female-headed households as the dominant way of disaggregating gender data often overlooks the vast majority of women who reside in male-headed households, and only gives a partial view of gender inequalities (Mehra and Rojas 2008). It limits the understanding of the complexity of gender issues and often tends to exaggerate gender inequalities and the asset poverty of women.

In the African context of small-scale mixed crop-livestock systems, men and women have different preferences regarding livestock and livestock products, and engage in many livestock production, management and marketing activities. Gender analysis should, therefore, be conducted beyond the simple differentiation of men and women, and, more specifically, female-headed households and male-headed households. The studies reported in this book suggest that gender analysis in livestock should use multiple data collection methods by both male and female enumerators, interviewing men and women within the same household as individual and joint decision-makers, and asking about individual preferences, ownership, access and control, as well as decision-making on livestock and livestock products marketing. This approach is in line with some of the best practices in gender analysis that require that ownership and decision-making should not be conflated, and always asking about the ownership of assets at the individual level, while allowing for the fact that assets may be jointly owned by a couple or more than one owner (Deere *et al.* 2012). It provides a more rigorous and nuanced understanding of the complexity of gender dynamics in livestock. Recognizing this complexity, gender analysis in agriculture has seen the development and field testing of new tools and interesting survey instruments for collecting gender disaggregated data. These include *The Women's Empowerment in Agriculture Index* (Alkire *et al.* 2012), the gender assets profile gaps (Deere *et al.* 2012; Doss *et al.* 2008), the gender mapping (Meinzen-Dick *et al.* 2012) and the gender transformative approach (CGIAR 2012). This book is a useful addition to these analyses.

These tools are now field tested and adapted in livestock research and development projects in different contexts. The extent to which these new tools will capture the complexity of gender and intra-household decision-making in livestock production and marketing systems is yet to be seen. Intra-household decision-making is dynamic, complex and evolving in different contexts and opportunities. Decision-making processes are not easily captured through surveys, and may require more ethnographic approaches.

Who owns and controls livestock? Who decides? Who benefits?

In livestock systems, the concept of ownership cannot be taken in isolation from control of and decision-making over benefits. It should not be assumed that the individual who owns the livestock necessarily has all the decision-making powers on access, use and control of all the benefits. Results presented in this book show considerable nuances in terms of ownership and benefits from livestock. For example, while women may not own dairy cows, they enjoyed some autonomy in the sale of milk from these cows. Another difference demonstrated was that women derived more benefits from exotic chickens than indigenous chickens and from dairy cows more than indigenous cows, mainly due to the potential market for milk and eggs from these exotic species. Women derived more benefits from these species owing to their joint ownership of these species with men. Typically, women owned more indigenous species, especially chickens, alone and more exotic species jointly with men. Women were also more actively involved in the marketing of livestock products like eggs and milk than in the marketing of livestock such as chickens, small stock and cattle.

One significant finding from the studies reported in this book was the extent of joint decision-making within farm households than was previously acknowledged. Whereas previous studies have shown that households do not act in a unitary manner when making decisions or allocating resources, and that men and women within households do not always have the same preferences, nor do they always pool their resources (for a review see Quisumbing 2003), empirical results from this study show that joint ownership of livestock and joint decision-making are more common in the largely mixed crop-livestock systems of Kenya and Tanzania. For example, more animals are owned jointly by men and women than by women alone, and even when men own the livestock, women derive benefits from these livestock irrespective of whether they co-own them or not. There is no conclusive evidence as to whether women's sole ownership of livestock has more beneficial outcomes than joint ownership and joint decision-making, as is indicated by the statement that "what matters are a woman's own income and assets . . . all of which increase her bargaining power and ability to influence household choices" (World Bank 2012: 21) and draws attention to the importance of context when reaching conclusions. It is conventionally accepted that when crops or livestock are produced to generate income, men often take over the decision-making matters related to the sale of animals and animal products, and the distribution of income benefits within the household (FAO 2011). The *World Development Report 2012* on gender equality (World Bank 2012) reports that a significant proportion of women do not make decisions even on their own income and assets. This is the case in many contexts, and the need for women to own their own assets cannot be over-emphasized. Nevertheless, in situations where women benefit from jointly owned assets with men, or assets owned by men alone, the ability to access, control and make decisions over these assets and benefits should not be ignored but should be promoted in tandem with the agenda to promote asset ownership by women.

Based on sex-disaggregated data collected from men and women within the same households on individual sources of income, previous chapters in this book analyse gender differences in decision-making and management of income across livestock and livestock products. Studies reviewed by Kristjanson *et al.* (2010) demonstrated a marked pattern of gender differences in control of income based not only on the type of livestock but on the species. In general, livestock and livestock products with a regular flow of small income are controlled by women, whereas men control income from large livestock sales. In Kenya, however, there was no difference in the proportion of income from sale of large and small stock managed by women. Most income from livestock and livestock products was jointly managed. Women managed income from sales of milk and eggs even when they did not own the dairy cows or the exotic chickens. Milk was an important source of income, contributing up to 40 per cent of all the livestock income and 29 per cent of all the household income. The study finds that women's income management is influenced by the amount of income going into the household. Their decision-making power and control of income increased significantly when they owned other assets, or when the households had multiple sources of income (Njuki *et al.* 2011).

The studies reported in this book nuance earlier findings that men sell women's livestock and assets and control the income from the sales, and that a significant proportion of women do not make decisions even about their own income and assets (Quisumbing *et al.* 2012; World Bank 2012). The studies show that men do not necessarily assert their control over "female" livestock and livestock products that have become lucrative. Women are able to market chickens, eggs and milk on their own and, if men market them, income accrued from these sales is shared with women and, hence, benefits the household. When men sold eggs, women retained 60 per cent of income share. A large proportion of women could still not make decisions on their own livestock, however, and had to consult their husbands. Further research that looks at species ownership alongside benefits that women get from these species would be useful, as women may own fewer of a particular species but derive more benefits from that species than another species where they own more. It is possible that joint ownership and joint decision-making can both increase food security and be transformative, making intra-household relations more productive and empowering women as a result (Farnworth 2012).

Upgrading women's livestock value chains

In order to be successful, livestock programs have to address the multiple market failures and constraints that limit women's participation in livestock value chains. Improving market access has become one of the most strategic pillars of any agricultural program (Fischer and Qaim 2012). It is recognized that women face several constraints in participating in markets and have far less access to higher-value markets, and their crops and livestock may be sold on their behalf by men, who often keep and control the income (Dolan 2001; World Bank 2012). This book shows that women do not participate in marketing of large animals but dominate marketing of local chickens and animal products, often in informal markets at farm gate. A central question, therefore, would be to investigate what productive and marketing strategies women can adopt in order to successfully upgrade and benefit from livestock value chains. Can these strategies lead to women's empowerment and control, and more efficient use of income by women? How can livestock programs and projects design, test and promote innovations for engendering livestock value chains and upgrading women's livestock value chains?

Engendering value chains and upgrading women's livestock value chains present opportunities to narrow gender gaps in livestock ownership and decision-making (KIT *et al.* 2012). Upgrading is a key concept in value chain analysis that refers to the process of acquiring technological, institutional and market capabilities that allow firms (individuals or communities) to improve their competitiveness and move into higher-value activities. It is the desirable change in value chain participation that increases rewards and/or reduces exposure to financial and other risks associated with poverty, gender and the environment (Bolwig *et al.* 2011). For KIT *et al.* (2012), upgrading women's value chains entails finding ways to remove gender inequalities in value chains and empowering women to expand their capabilities and opportunities to create value and control this value, and obtain better returns from their livestock and livestock products. It means improving the performance of livestock value chains and making them work for women and benefit women and their households.

Drawing from a dozen empirical case studies with a range of livestock products, KIT *et al.* (2012) describe practical strategies, approaches and tools for engendering value chains and improving the performance of value chains to ensure that women can participate in and benefit from upgrading value chains. These include: (i) working with men on typical livestock products controlled by women such as chickens, eggs and milk; (ii) opening up opportunities for women to work on what are considered to be "men's" livestock and livestock products and markets, such as cattle and formal cooperatives; (iii) building women's capacity, organization, sensitization and access to finance and information; (iv) using standards and certification to promote gender equity; and (v) promoting gender responsible business.

One successful strategy for upgrading women's livestock value chains is investing in and strengthening women's social capital. As the chapters in this book demonstrate, women's groups and similar forms of collective marketing can contribute to increasing women's bargaining position, enable women to access high-value markets and reduce transaction costs. Women's organizations and marketing collectives such as cooperatives serve multiple purposes that are often beneficial to women. They play important roles in accelerating adoption of technologies, accessing market information and credit, and building financial and social assets (Mayoux 2001; Quisumbing *et al.* 2012). Benefits for women notwithstanding, many studies found that men tend to dominate cooperative membership (Abebaw and Haile 2013). There is, therefore, a need to better understand under what conditions and through what mechanisms and forms of collective action women can benefit from livestock value chains, and exploit those that provide the greatest opportunities for women to benefit.

Does women's ownership and control of livestock improve food security?

The link between livestock and food security and nutrition is complex. There are at least four pathways through which livestock contribute to food security: (i) enabling direct access to animal source foods; (ii) providing cash income from sale of livestock and livestock products that can in turn be used to purchase food, especially during times of food deficit; (iii) contributing to increased aggregate food supply as a result of improved productivity from use of manure and traction; and (iv) lowering prices of livestock products and, therefore, increasing access to such products by the poor, especially poor urban consumers through increasing livestock production.

Results on the food security impacts of livestock ownership are mixed. The study findings demonstrate a clear pattern of positive relationships between women's ownership of livestock and control of income and two of the three measures of food security: consumption of some animal source foods and household diet diversity scores. In contrast, the findings show a negative relationship between food availability as measured by the Months of Adequate Food Supply within the household with women's ownership of livestock and control of income. This contradiction may be explained, in part, by the fact that chickens, often owned and controlled by women, seem to contribute more to household food security in terms of both the consumption of animal source foods and the household dietary diversity scores. This is, however, only true for exotic and not indigenous chickens. Generally, households tend to keep just a few indigenous chickens, whose production of eggs is neither prolific nor regular. Exotic chickens produce more eggs, which are easily and regularly consumed and sold to provide a small but regular income controlled by women and that women often use to purchase food.

The results on food security impacts are not as conclusive for milk as they are for exotic chickens. The frequency of milk consumption was significantly lower in households where women owned cattle compared to households where women did not own cattle. These results suggest that livestock ownership does not necessarily result in an increase in the consumption of animal source food and the diversity of diets. There seems to be an inverse relationship between livestock ownership and consumption of meat and milk; the frequency of meat consumption was higher in households without cattle, goats and exotic chickens than households that did own them. There was also no evidence that income from sale of livestock and livestock products was used to purchase food. These findings concur with earlier findings by Nicholson *et al.* (1999), who found that dairy cattle ownership does not always translate to an equivalent improvement in nutritional outcomes. This could be because dairy incomes are mainly controlled by men and purchase of food is rarely a priority expenditure item for men.

There is now a renewed attention to exploring and testing promising pathways for "nutrition-sensitive" agriculture and food security, and for explicitly outlining the "pathways of change" from agricultural production and marketing to consumption and improved food security and nutrition. These pathways are far from being linear and simple, but are nebulous and complex (for a review, see Fan and Pandya-Lorch 2012; Masset *et al.* 2011). Clearly, the results documented in this book are more exploratory than conclusive, and call for more rigorous research on the links between livestock ownership and livestock value chains, and their implications for food and nutrition security.

Engendering livestock research for development

There is no lack of policies, strategies, frameworks, guidelines and tools for mainstreaming gender in research and development. However, progress has been slow and important challenges remain. Recent reviews of experiences and lessons in integrating gender in agricultural research have identified several factors that limit effective integration of gender and delivery of scientific and development outcomes in international agricultural research centres (CGIAR 2012; ILRI 2012b; Meinzen-Dick *et al.* 2010). These reviews have also recommended a set of enabling changes that are needed to translate gender frameworks into actual research and development practice. While intentions are often well expressed in proposals and strategic plans and gender strategies, effective integration of gender in livestock research is often fraught with a number of challenges.

The first challenge is conceptual and results from the failure of theory to frame a systemic gender theory of change and an impact pathway in livestock research and development. Gender is often framed and used in instrumental terms to improve the conditions of women on a case by case basis, but there is less consideration of how to improve the position of women overall, and influence strategic gender relations. This raises the challenge of making gender a respected field of research, capable of generating high-quality publications and not only anecdotal stories. Addressing this challenge would make gender less of a "common-sense" and amateur practice of any scientist. There are several recommendations in the Stripe Review of social sciences in the CGIAR (Consultative Group on International Agricultural Research) that provide insight into how to make gender a respected field in the social sciences (CGIAR 2009).

The second challenge is frequent implementation failures. A persistent problem in agricultural research organizations, and particularly in livestock research organizations, relates to insufficient core capacity and funding for gender research to implement well-meaning gender strategies and frameworks, and to move beyond gender analysis to effective transformation approaches for empowering women and the rural poor. To deliver on gender outcomes, research organizations will have to acquire and continue to develop high-quality teams of gender experts and social scientists with sufficiently diverse disciplinary skills and an ability to provide scientific leadership on the design and other key social science issues during the different stages of agricultural research. This will create a broader focus on a clear vision of change and will help to move away from the gender tool-kit approaches, which have now revealed their limitations. It is also important to identify the factors that have limited effective integration and delivery of gender outcomes in livestock research and learn from successful approaches in other sectors.

There are now renewed efforts to integrate some gender transformative approaches (CGIAR 2012) in agriculture and livestock research. Figure 9.1 gives an overview of the gender transformative approach and its defining core characteristics. Gender transformative approaches go beyond gender analysis to address some of the social norms, attitudes and behaviours, power relations and social systems that underlie and entrench gender inequalities. These approaches engage with the political dimensions of women's empowerment and require intensive efforts and resources to achieve change. Adopting a gender transformative approach in livestock research requires a clear systematic and coherent vision of change, and a strong commitment to solving important problems that impact women and other marginalized and often poor people. It requires new research models that promote a shift from very small, short-duration gender analysis projects that are tool-based and accommodative in vision and actions to much larger and longer-term projects that can experiment with more gender transformative gender approaches that empower social change. Many livestock research organizations do not have such resources. Nevertheless, opportunities for impact-oriented partnerships with development organizations that facilitate uptake of livestock research and its long-run impact are now emerging, indicating the need to be more strategic and to give priority to identifying and targeting shared outcomes and impacts where research on gender can make a critical contribution.

The third challenge relates to delivery of gender outcomes given the diversity of agriculture and livestock interventions, and impatience for results. A persistent challenge is reaching agreement on a few common elements that can be monitored and are significant in terms of their impact on gender across many projects and programs. This is the challenge of identifying and clarifying the "missing middle" or intermediate outcomes in relation to expected changes in gender inequality that are crucial for final impact. To overcome this challenge, it is important to focus on a few strategic research questions as a first step to avoid scattering efforts across the wide panorama of gender issues in livestock research and development. An important contribution of this book is to provide some illustrative examples of a few,

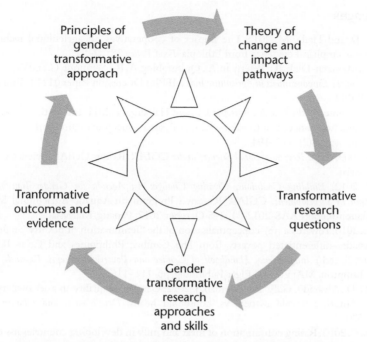

FIGURE 9.1 Prerequisite for a gender transformative approach in livestock research

Source: Adapted from CGIAR (2012).

well-chosen process indicators that could be defined collectively. It is important to recognize that livestock research projects and programs will often have limited and shorter-term objectives to strengthen implementation and delivery of gender outcomes. It is also much harder to work on the bigger picture of social change, or transforming gender relations than conducting the traditional and non-gender-transformative gender analyses.

These challenges are enormous and require critical systems thinking to (i) sharpen gender diagnosis into how livestock systems and target beneficiary groups are defined and targeted – getting away from the generic use of "women and men", "the poor, especially women" and "female-headed households"; (ii) identify some "core" shared variables, indicators and measurement instruments across projects that can be adapted to different contexts but used across sites and regions so that larger-scale studies and their databases can be developed in order to research the issue of who owns, manages and controls livestock and associated assets and benefits, and who makes decisions over them; (iii) experiment with foresight studies and scenario analyses with a gender focus to explore pos- sible futures of the livestock revolution; and (iv) figure out how to partner with gender transformative development programs that enable women to benefit from agricultural innovations.

References

Abebaw, D. and Haile, M. 2013. The impact of cooperatives on agricultural technology adoption: empirical evidence from Ethiopia. *Food Policy* 38: 82–91.

Alkire, S., Meinzen-Dick, R., Peterman, A., Quisumbing, A. R., Seymour, G. and Vaz, A. 2012. *The Women's Empowerment in Agriculture Index*. IFPRI Discussion Paper 01240. Washington, DC: IFPRI.

Bolwig, S., Ponte, S., du Toit, A., Riisgaard, L. and Halberg, N. 2011. Integrating poverty and environmental concerns into value-chain analysis: a conceptual framework. *Development Policy Review* 28(2): 173–194.

CGIAR 2009. *Stripe Review of Social Sciences in the CGIAR*. Rome: CGIAR Science Council Secretariat.

CGIAR 2012. *Building Coalitions, Creating Change: An Agenda for Gender Transformative Research in Development*. CGIAR Research Program on Aquatic Agricultural Systems Workshop Report AAS-2012-31, 3–5 October 2012, Penang, Malaysia.

Chant, S. 2010. Towards a (re)-conceptualisation of the "feminisation of poverty": reflections on gender-differentiated poverty from The Gambia, Philippines and Costa Rica. In Chant, S. (ed.) *International Handbook of Gender and Poverty: Concepts, Research, Policy*. Northampton, MA: Edward Elgar Publishing, pp. 111–116.

Deere, C. D., Alvarado, G. E. and Twyman, J. 2012. Gender inequality in asset ownership in Latin America: female owners vs. household heads. *Development and Change* 43(2): 505–530.

Delgado, C. 2003. Rising consumption of meat and milk in developing countries has created a new food revolution. *Journal of Nutrition* 133(11): 3907S–3910S.

Delgado, C., Rosegrant, M., Steinfeld. H., Ehui, S. and Courbois, C. 1999. *Livestock to 2020: The Next Food Evolution*. Food, Agriculture and the Environment Discussion Paper 28. Washington, DC: IFPRI.

Dolan, C. 2001. The "good wife": struggles over resources in the Kenyan horticultural sector. *Journal of Development Studies* 37(3): 39–70.

Doss, C., Grown, C. and Deere, C. D. 2008. *Gender and Asset Ownership: A Guide to Collecting Individual-level Data*. Policy Research Working Paper 4704. Washington, DC: World Bank. Available at: http://econ.worldbank.org/docsearch (accessed 28 December 2012).

Fan, S. and Pandya-Lorch, R. (eds) 2012. *Reshaping Agriculture for Nutrition and Health*. Washington, DC: IFPRI.

FAO 2011. *The State of Food and Agriculture 2010–11: Women in Agriculture – Closing the Gender Gap for Development*. Rome: FAO.

Farnworth, C. R. 2012. *Gender Approaches in Agricultural Programmes*. FAO Zambia country report. Rome: FAO.

Fischer, E. and Qaim, M. 2012. Linking smallholders to markets: determinants and impacts of farmer collective action in Kenya. *World Development* 40(6): 1255–1268.

ILRI 2012a. *ILRI Corporate Report 2010–2011. Livestock Matter(s): Where Livestock Can Make a Difference*. Nairobi: ILRI.

ILRI 2012b. *Strategy and Plan of Action to Mainstream Gender in ILRI*. Nairobi: ILRI.

KIT, Agri-ProFocus and IIRR 2012. *Challenging Chains to Change: Gender Equity in Agricultural Value Chain Development*. Amsterdam: KIT Publishers, Royal Tropical Institute.

Kristjanson, P., Waters-Bayer, A., Johnson, N., Tipilda, A., Njuki, J., Baltenweck, I. *et al.* 2010. *Livestock and Women's Livelihoods: A Review of the Recent Evidence*. ILRI Discussion Paper No. 20. Nairobi: ILRI.

Masset, E., Haddad, L., Cornelius, A. and Isaza-Castro, J. 2011. *A Systematic Review of Agricultural Interventions that Aim to Improve Nutritional Status of Children*. London: EPPI-Centre, Social Science Research Unit, Institute of Education, University of London.

Mayoux, L. 2001. Tackling the down side: social capital, women's empowerment and micro-finance in Cameroon. *Development and Change* 32(3): 435–464.

Mehra, R. and Rojas, M. H. 2008. *Women, Food Security and Agriculture in a Global Market Place: A Significant Shift*. International Center for Research on Women (ICRW). Available at: http://www.icrw.org/publications/women-food-security-and-agriculture-global-marketplace (accessed 9 December 2012).

Meinzen-Dick, R., Quisumbing, A., Behrman, J., Biermayr-Jenzano, P., Wilde, V., Noordeloos, M. *et al.* 2010. *Engendering Agricultural Research*. IFPRI Discussion Paper 973. Washington, DC: IFPRI. Available at: http://www.ifpri.org/publication/engendering-agricultural-research (accessed May 2013).

Meinzen-Dick, R., van Koppen, B., Behrman, J., Karelina, Z., Akamandisa, V., Hope, L. *et al.* 2012. *Putting Gender on the Map: Methods for Mapping Gendered Farm Management Systems in Sub-Saharan Africa*. IFPRI Discussion Paper 01153. Washington, DC: IFPRI.

Mupawaenda, A. C., Chawatama, S. and Muvavarirwa, P. 2009. Gender issues in livestock production: a case study of Zimbabwe. *Tropical Animal Health and Production* 41:1017–1021.

Nicholson, C. F., Thornton, P. K., Mohammed, L., Minge, R. W., Mwamwchi, D. M., Elbasha, E. H. *et al.* 1999. *Smallholder Dairy Technology in Coastal Kenya: An Adoption and Impact Study*. Nairobi: ILRI.

Njuki, J., Kaaria, S., Chamunorwa, A. and Chiuri, W. 2011. Linking smallholder farmers to markets, gender and intra-household dynamics: does the choice of commodity matter? *European Journal of Development Research* 236: 426–443.

Pica-Ciamarra, U. and Otte, J. 2009. *The 'Livestock Revolution': Rhetoric and Reality*. Pro-Poor Livestock Policy Initiative, a Living from Livestock Research Report. Available at: http://www.fao.org/ag/AGAinfo/programmes/en/pplpi/docarc/rep-0905_livestockrevolution.pdf (accessed May 2013).

Quisumbing, A. R. (ed.) 2003. *Household Decisions, Gender, and Development: A Synthesis of Recent Research*. Washington, DC: IFPRI.

Quisumbing, A. R. and Meinzen-Dick, R. 2001. *Empowering Women to Achieve Food Security*. 2020 Focus 6, Policy Brief 1 of 12. Washington, DC: IFPRI. Available at: http://ageconsearch.umn.edu/bitstream/16032/1/vf010006.pdf (accessed 8 January 2012).

Quisumbing, A. R., Payongayong, E., Aidoo, J. B. and Otsuka, K. 2001. Women's land rights in the transition to individualized ownership: implications for the management of tree resources in western Ghana. *Economic Development and Cultural Change* 50(1): 157–182.

Quisumbing, A., Meinzen-Dick, R., Raney, T., Croppenstedt, A., Behrman, J. A. and Peterman, A. (eds) forthcoming. *Gender in Agriculture and Food Security: Closing the Knowledge Gap*. Dordrecht and Rome: Springer and FAO.

Saghir, P., Njuki, J., Waithanji, E., Kariuki, J. and Sikira, A. 2012. *Integrating Improved Goat Breeds with New Varieties of Sweet Potatoes and Cassava in the Agro-pastoral Systems of Tanzania: A Gendered Analysis*. ILRI Discussion Paper 21. Nairobi: ILRI.

Staal, S., Poole, J., Baltenweck, I., Mwacharo, J., Notenbaert, A., Randolph, T. *et al.* 2009. *Strategic Investment in Livestock Development as a Vehicle for Rural Livelihoods*. ILRI Knowledge Generation Project Report. Nairobi: ILRI.

UNDP 2012. *Africa Human Development Report 2012: Towards a Food Secure Future*. New York: UNDP. Available at: http://www.undp.org/content/dam/undp/library/corporate/HDR/Africa%20HDR.pdf (accessed May 2013).

Wooten, S. 2003. Losing ground: gender relations, commercial horticulture, and threats to local plant diversity in rural Mali. In Howard, P. L. (ed.) *Women and Plants: Gender Relations in Biodiversity Management and Conservation*. London: Zed Books.

World Bank 2012. *World Development Report 2012: Gender Equality and Development*. Washington, DC: World Bank.

World Bank, FAO and IFAD 2009. *Gender in Agriculture Sourcebook*. Washington, DC: World Bank.

INDEX

References in **bold** indicate tables and in *italics* indicate figures.

Printed and bound by CPI Group (UK) Ltd, Croydon, CR0 4YY

21/10/2024

01777044-0018